HOW TO USE
WOODWORKING TOOLS
EFFECTIVELY AND SAFELY

BARNES LARGE TYPE EDITIONS

ALSO BY VERNON M. ALBERS

HOW TO USE WOODWORKING TOOLS EFFECTIVELY AND SAFELY

by
Vernon M. Albers

SOUTH BRUNSWICK AND NEW YORK:
A. S. BARNES AND COMPANY
LONDON: THOMAS YOSELOFF LTD

© 1975 by A. S. Barnes and Co., Inc.

Library of Congress Catalogue Card Number: 74-9277

A. S. Barnes and Co., Inc.
Cranbury, New Jersey 08512

Thomas Yoseloff Ltd
108 New Bond Street
London W1Y OQX, England

First Printing of Large Type Edition 1976

ISBN 0-498-01851-2 (Large Type Edition)

PRINTED IN THE UNITED STATES OF AMERICA

CONTENTS

PREFACE

Wood is a marvelous material. It has great strength for its weight and it has great beauty if it is properly used. It also has great durability, as evidenced by the large number of pieces of furniture which are still in use after more than 100 years. No two pieces of wood are exactly alike and no one piece is uniform. The characteristics of the wood in the direction of the grain are quite different from those across the grain because wood is made up of fibers which run in the direction of the grain. Because of the nonhomogeneous nature of wood, more skill is required in working with it than is required in working with a homogeneous material such as plastic or metal.

Special tools are required for woodworking and any cutting tools must be very sharp if they are to do their job properly. It is important to keep all tools properly sharpened and adjusted so that they will always be ready for use.

In my previous books on woodworking I have concentrated on the methods of construction or repair of representative types of furniture or cabinets. It was necessary to describe, to some extent, the appropriate tools for the processes described. However, the primary focus was on the construction or repair processes, usually indicating the simplest tools which could be used to do the job.

In this book I will attempt to show how an amateur can most effectively use and care for all of the various tools which he might have in his workshop.

Although power tools are convenient for many operations, there are also many operations which must be done with hand tools; other jobs, which could be done with power tools, may be more effectively done with hand tools if the operator has adequate skill and the tool is in proper condition. There are also many instances where certain hand tools, such as measuring and layout tools, must be used, even though the cutting operations are done with power tools.

For our purpose, we will define a hand tool as any tool which does not use supplementary power in its operation. Hand electric drills and hand power saws, for example, will be treated as power tools.

Woodworking tools can be hazardous, so procedures for the safe operation of the tools will be emphasized.

HOW TO USE
WOODWORKING TOOLS
EFFECTIVELY AND SAFELY

1

MEASUREMENT, ROUGH CUTTING AND LAYOUT

1.1 INTRODUCTION

Before cutting any lumber it is necessary to lay out the pieces to be cut. The layout requires the use of linear measuring instruments, angle measuring instruments, squares, compasses, dividers and curved surfaces such as French curves or flexible curves.

A sample of the various instruments is shown in Fig. 1.1:

1. *A* is a steel square. The length of the long arm of the square is 24" and that of the short arm is 12". The scales on the square are convenient for measuring lengths up to 24" and it is used for squaring large pieces such as pieces of plywood. The square is a precision instrument so it may also be used as a straight edge.

2. *B* is a try square, which is used, primarily, as a square for drawing lines perpendicular to the edge of a piece of lumber and for checking the square-ness of a piece already formed. The handle of the try square is thicker than the blade to facilitate this. Although the blade of a try square has a scale ruled

along its edge, its use for measurement is of second-
ary importance. The portion of the blade extending from the handle is usually 7½" to 8" long.

3. A combination square is shown at *C*. It consists of a
12" blade with scales on the two sides of each edge.
One scale is divided into ⅛" sections, two of the
scales are divided into 1/16" sections and one is divided into 1/32" sections. The blade is removable
from the sliding "T," which can be clamped at any

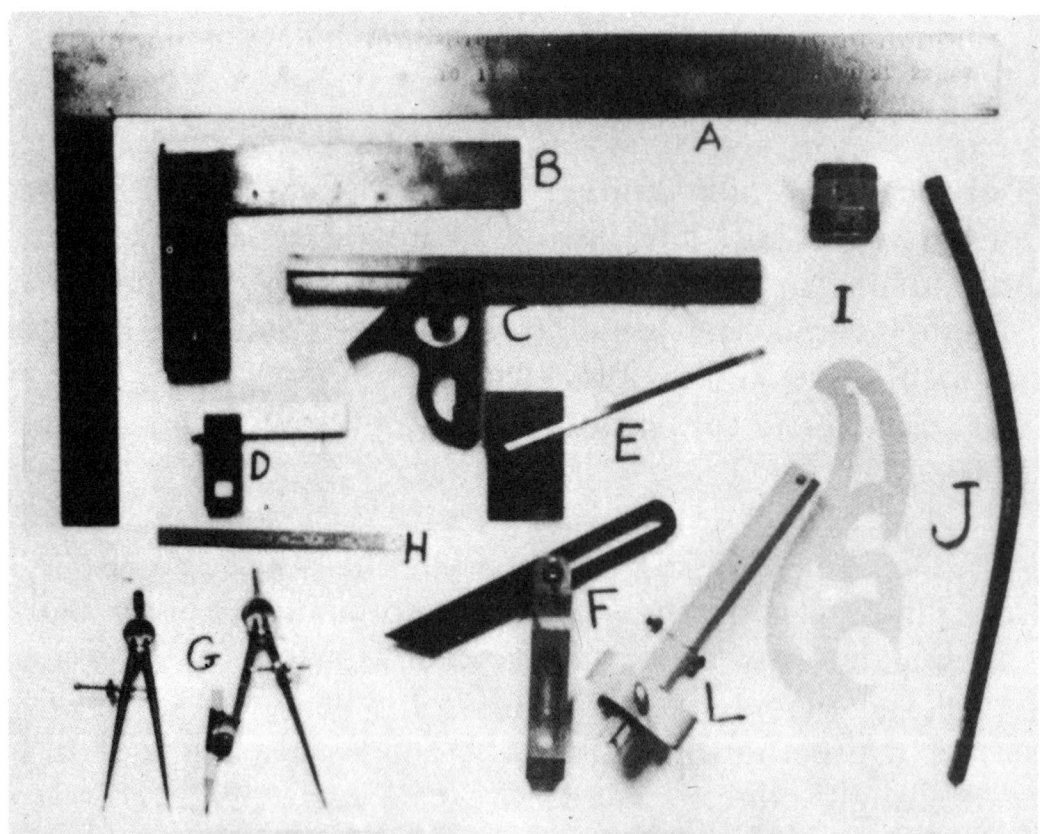

Fig. 1.1 **The various tools used in layout.** *A* **is the large
steel square,** *B* **is the try square,** *C* **is the combination square,**
D **is the small square,** *E* **is the protractor,** *F* **is the sliding
"T" bevel,** *G* **is the compass and divider,** *H* **is the small steel
scale,** *I* **is the steel tape,** *J* **is the French curve and flexible
curve and** *L* **is the marking gauge.**

position along the blade. One side of the sliding "T" makes an angle of 90° with the blade and the other side makes an angle of 45° with the blade. When the blade is removed from the sliding "T," it serves as a precision one-foot scale or straight edge.

4. The small square shown at *D* is similar to the combination square except that the blade is only four inches long and it has scales divided in eighths, sixteenths, thirty-seconds and sixty-fourths of an inch. The "T" has only a 90° angle but it can be clamped at any position along the blade. This square is particularly convenient for checking the squareness of small pieces and the blade, when removed from the "T," may be used as a precision scale and straight edge.

5. The protractor shown at *E* is used for laying off angles other than 45° and 90°. A good steel protractor can be read to about 1/2°.

6. A sliding "T" bevel is shown at *F*. This tool is similar to the try square except that the angle of the blade is adjustable and the position of the "T" can be adjusted along the slot in the blade. There is no angle scale on the sliding "T" bevel but its angle can be set by using the protractor.

7. The instruments at *G* are the compass and divider. The divider has two steel points and the compass has a steel point on one arm and a pencil point on the other. The divider is used for laying out equal spacing of points and the compass is used for drawing small circles about a fixed center.

8. *H* is a small, thin steel scale, usually six inches long, with scales divided into thirty-seconds and sixty-fourths of an inch on the front side. The back of the scale has tables for converting common fractions of an inch to thousandths of an inch.

9. The flexible steel tape at *I* is used for measuring long lengths.
10. The French curve and flexible curve at *J* are used for laying out curved lines.
11. *L* is a marking gauge used for marking lines parallel to an edge of a piece of stock.

1.2 ROUGH CUTTING LUMBER STOCK

When you start to contruct a piece of furniture, you should first examine your available lumber and select the pieces for the various parts. The selection will depend on the kind of furniture you are about to construct. For example, the most prominent portion of a chest of drawers is the drawer fronts. The pieces to be cut for the drawer fronts should be selected to have similar grain and the individual drawer fronts should be chosen so that the orientation of the grains on the different drawer fronts will be similar.

If the piece is to have a glued-up top or panel, the lumber should be examined to determine if the pieces, in the rough, have any warp or twist. It is important that the resulting top or panel be as flat as possible. This can be achieved by selecting the pieces to be cut and orienting them so that warp or twist in adjacent pieces will compensate.

When you have decided how you are going to cut the various parts, the pieces are layed out on the rough stock with the steel tape, allowing at least 1" in length for final trimming, and the lines of the cuts are made perpendicular to the edge by means of the try square. The ends of the rough stock often have cracks extending a few inches into the board and this portion should be cut off as waste.

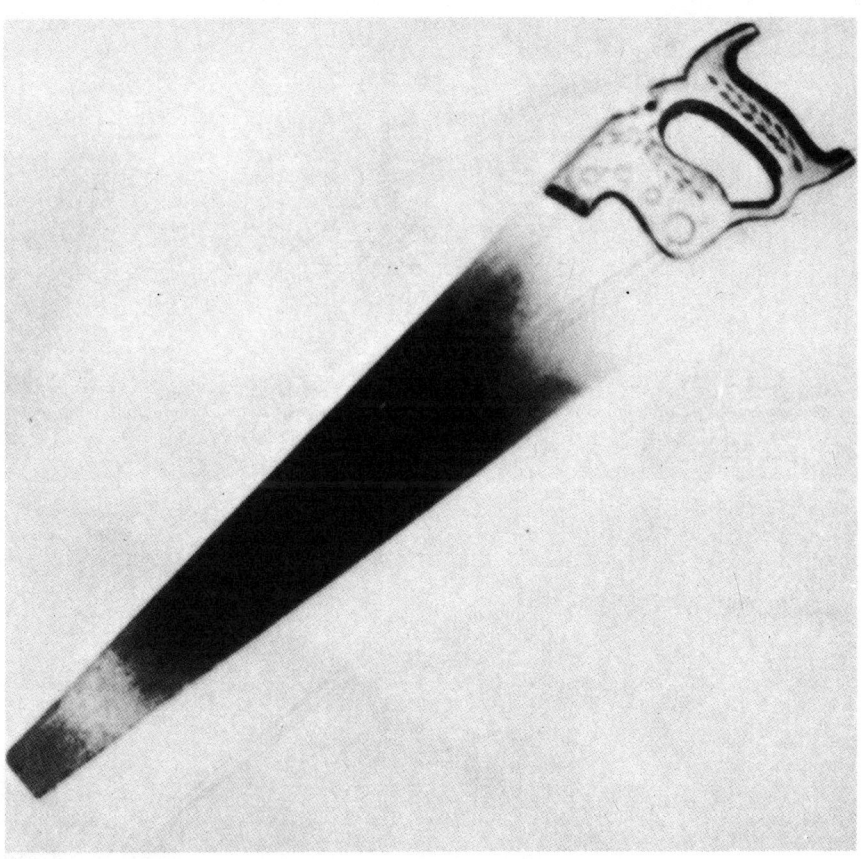

Fig. **1.2** **A hand crosscut saw.**

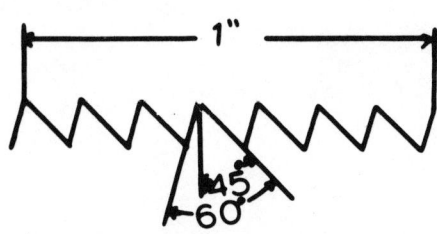

Fig. **1.3** **Arrangement of teeth on a crosscut saw.**

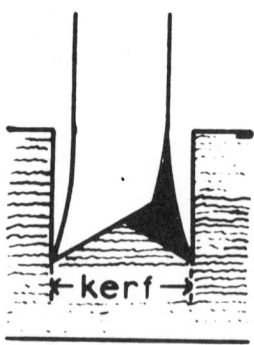

Fig. **1.4** **How the teeth on a hand crosscut saw cut.**

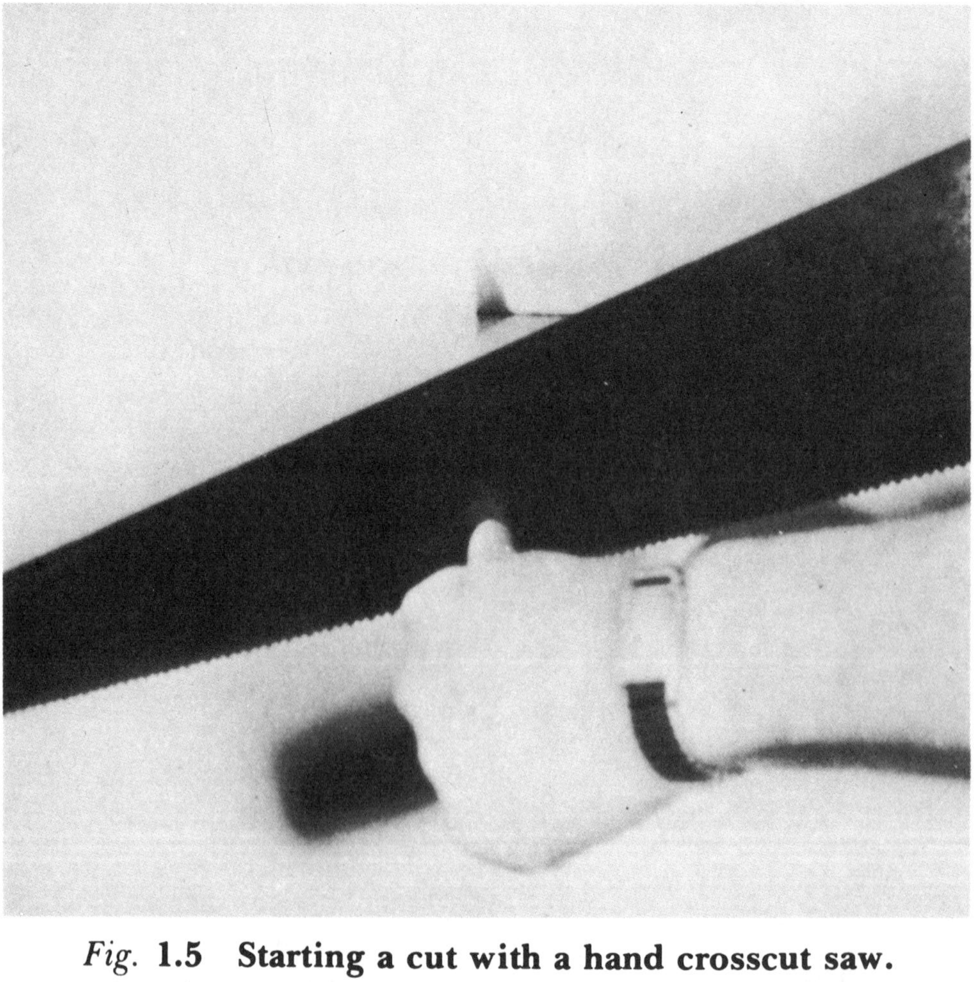

Fig. **1.5** **Starting a cut with a hand crosscut saw.**

The rough lumber stock often is supplied in long lengths of 12', 14' or 16'. The rough cutting can often be done more conveniently with a hand saw, even though a power saw is available. An 8-point saw is usually used for this. Figure 1.2 is a photograph of a typical hand crosscut saw. Figure 1.3 shows the arrangement of teeth on an 8-point crosscut saw and Fig. 1.4 shows the cutting action of the teeth. Figure 1.5 is a photograph showing how a cut with the crosscut saw is started. Note the use of the thumb on the left hand to guide the position of the saw. The cut may be made with the line drawn on the stock straddled by the saw cut or the cut may be made along one side or the other of the line. The choice is determined by the way the original layout was made.

1.3   LAYOUT OF SQUARE AND ANGLE CUTS

Before square or angle cuts can be laid out, it is necessary to have one straight edge to work from. If lumber milled on the edges was purchased, the edges will be straight. However, rough lumber, which has been milled only on the sides, may not have straight edges. If the lumber does not have a straight edge, or if there are defects in the edge, it may be necessary to rip off a narrow portion of the edge. Figure 1.6 shows how a ripsaw cuts. After the rip cut is made the edge should be planed with a hand plane or a jointer to provide a smooth edge. The use of the hand plane and jointer will be described in Chapters 2 and 4.

A line at right angles to the edge may be laid out with the small square, the try square, the combination square or the steel square. The square used will depend on the

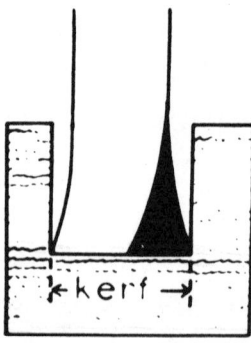

Fig. **1.6   How the teeth on a hand rip saw cut.**

Fig. **1.7   Scribing a line with a knife.**

size of the piece. In many instances, a fine pointed pencil will be adequate for marking but, if high precision is required, a very fine line can be scribed with a knife point, as indicated in Fig. 1.7.

The most common angle cut is made at an angle of 45°, since this is the angle at which miters are cut. The combination square is provided with a "T" which can be used for scribing either 90° or 45° angles.

The sliding "T" bevel can be set at any predetermined angle. Figure 1.8 shows the method of setting the sliding "T" bevel with a protractor. The angle can be set with

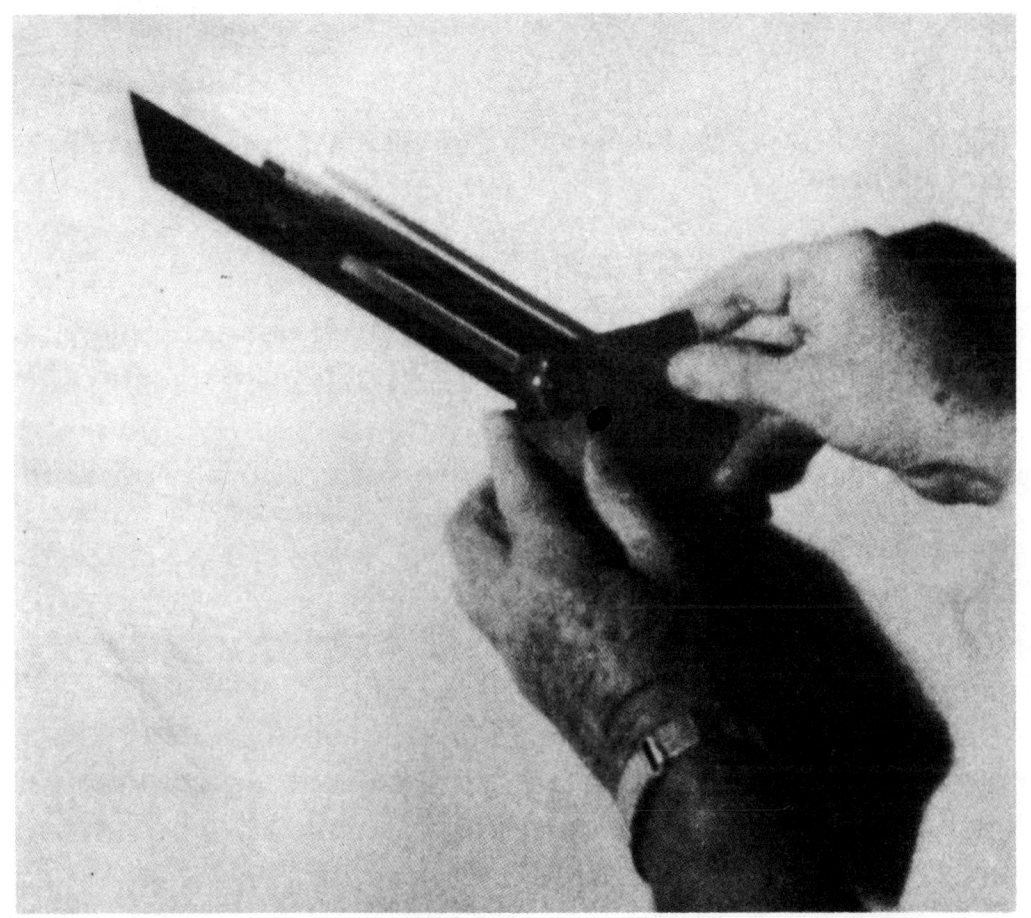

Fig. **1.8 Setting a sliding "T" bevel with a protractor.**

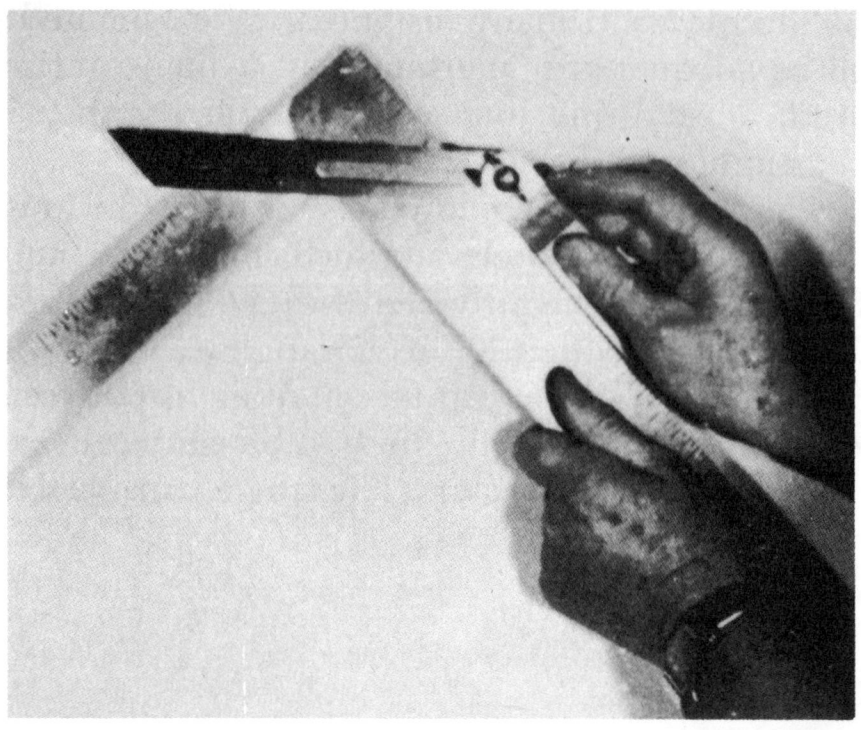

Fig. **1.9 Setting a sliding "T" bevel with the scales on a steel square.**

an accuracy of about ½° by this method. An alternate method of setting the sliding "T" bevel to 45°, using the scales on the steel square, is shown in Fig. 1.9. The bevel is set so that it intersects the same scale division on both scales.

1.4 LAYOUT OF CIRCLES AND IRREGULAR CURVES

Small circles are drawn with the compass and large circular arcs are scribed by use of a fine wire anchored at the center of the arc with a pencil attached to the scribing end of the wire. A wire should be used, rather than a string, because a string will stretch so the length of the radius will vary with the tension in the string. I prefer to make complicated curved layouts on heavy paper or poster board. The pattern can then be cut out and later traced on the wood. Usually, such a layout does not need to be made to absolute precision and your eye is the best judge of the desired shape. To obtain smooth transition between different parts of a curved layout, a French curve or a flexible curve can be used.

When a unit, such as a skirt on a cabinet, is laid out with a curved edge, it is desirable to lay out the portion from one end to the center on the pattern. To transfer the layout to the lumber, use the same pattern for both ends. It will, of course, have to be inverted for tracing the pattern on one end. This procedure insures that the two halves will be symmetrical.

1.5 LAYOUTS ON PLYWOOD

Plywood, especially hardwood plywood, is quite expensive, so the parts to be cut should be layed out with great care to minimize waste. You should, of course, take into account the desired direction of the grain in the various pieces.

It is desirable to use double lines between adjacent pieces to be cut from the plywood panel. One line is the finish line on one piece while the other is the finish line on the adjacent piece. You can then saw between the lines, which should be separated by an amount slightly greater than the width of the saw cut. A crosscut saw, 10-point or finer, should be used and the cutting should be done with the panel mounted on sawhorses. (See *Amateur Cabinetmaking* by Vernon M. Albers, A. S. Barnes and Co.)

1.6 LAYOUT OF JOINTS

Joints require layouts of high precision, since the strength of the joint is dependent on the perfection of the fit.

A mortise and tenon joint requires the formation of a rectangular slot, called the mortise, in the side of one piece and the cutting of the end of another piece to a size and length which will fit the mortise. A common form of mortise and tenon joint, shown in Fig. 1.10, requires that the face *a* of part *A* be flush with face *b* of part *B*. This requires that the shoulder between the tenon on *B* and the face *b* be equal to the width of the shoulder between the mortise and the face *a*. This layout

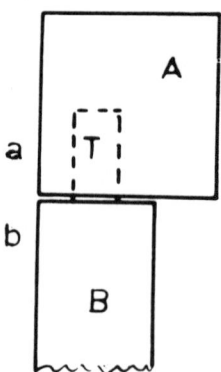

Fig. **1.10** **A mortise and tenon joint.**

can most easily be accomplished by use of the marking gauge. Since the position of the "T" on the marking gauge is movable on the bar, the bar can be extended beyond the "T" by an amount equal to the width of the shoulder, and the edge of the mortise is marked as indicated in Fig. 1.11a and the edge of the tenon marked as indicated in Fig. 1.11b. The bar is then extended by an

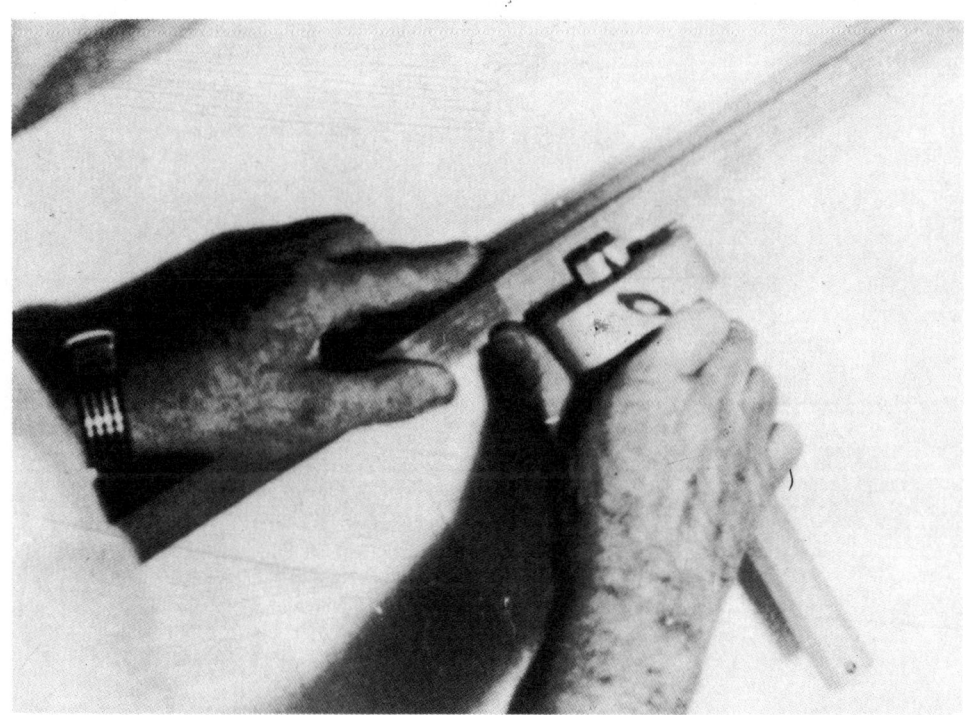

Fig. **1.11   Use of a marking gauge for layout of the parts of a mortise and tenon joint,** *a* **marking the tenon,** *b* **marking the edge of the mortise.**

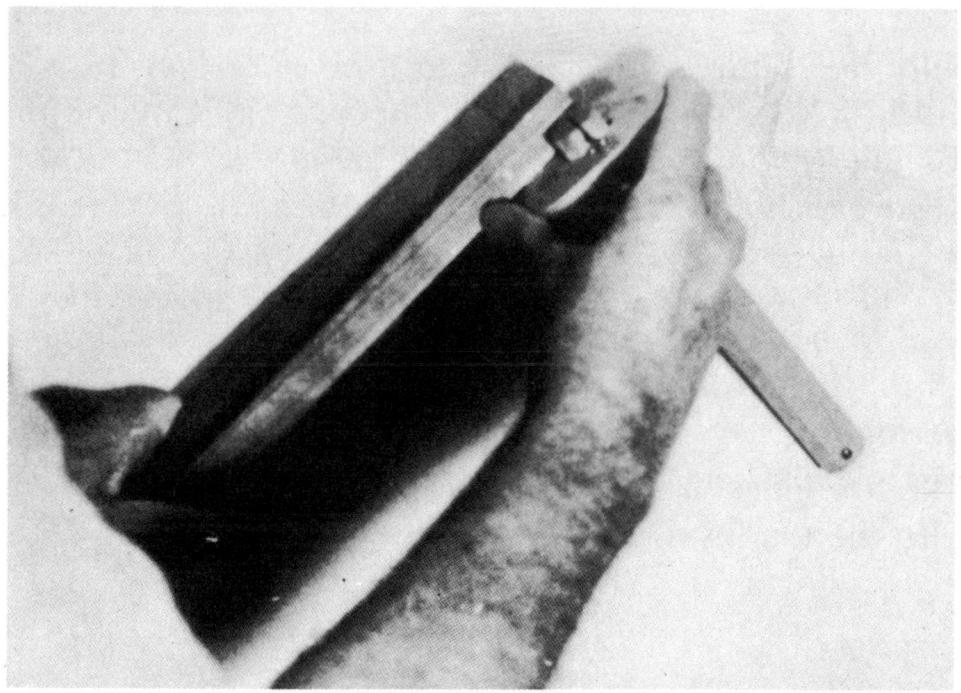

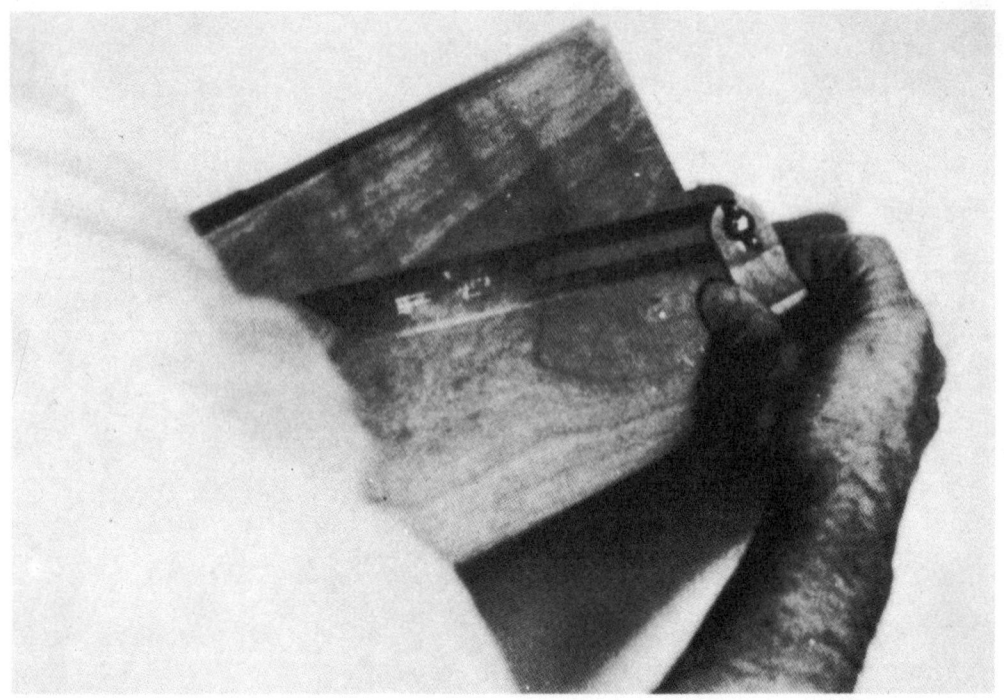

Fig. **1.12   Use of the sliding "T" bevel to lay out a dovetail joint.**

amount equal to the width of the shoulder plus the width of the mortise and the opposite sides of the mortise and tenon are marked in the same manner.

Dovetail joints are usually made with the dovetails cut at an angle of 15°, as indicated in Fig. 1.12. The sliding "T" bevel is set to the angle of 15°. This angle does not need to be set to high precision. The line indicating the depth of the tenons must be scribed before the edges of the tenons are marked. The edges of the dovetail mortises are marked after the tenons are cut, using the piece with the tenons as a pattern.

2
HAND CUTTING TOOLS

2.1 INTRODUCTION

Hand cutting tools are necessary and useful even though power tools are available. There are some operations which can be done with power tools but can be done more quickly with hand tools because of the time required to make the necessary set up to use the power tool. There are other operations which can be done better with hand tools than with the power tools available in the amateur's shop. There are some amateurs who do not have power tools or have a limited number of power tools available and, therefore, will need to do all or a part of the work with hand tools. All operations in furniture making can be done with hand tools, but some operations require more skill when done with hand tools.

2.2 SAWS

The large hand crosscut and rip saws have been described in Section 1.2. Regardless of the power tools available, a crosscut saw of eight to ten points per inch, similar to the one shown in Fig. 1.2, should be available in the amateur's shop. If a power saw is available a rip saw is not essential. A crosscut saw should be used for hand cutting plywood and, in the occasional instances when a piece of lumber must be ripped by hand, it is possible to use the crosscut saw. It is not possible to use a rip saw for cutting across the grain. However, if no power saw is available, both the hand crosscut and rip saws should be available.

Figure 1.4 and 1.6 show how the teeth on crosscut and rip saws cut. In cutting across the grain, the sharp edges of the teeth of a crosscut saw must cut the fibers at the edges of the kerf and the smoothness of the cut depends on the sharpness of the teeth and the fineness of the teeth. In cutting parallel to the grain (i.e., ripping) the teeth of the rip saw continuously cut off the ends of the grain fibers as the saw works its way through the cut.

Figure 2.1 is a photograph showing the back saw, the dovetail cutting saw and the scroll saw.

When accurate cuts are to be made, particularly on small pieces, a back saw should be used. A back saw has a steel beam along the back of the blade, to stiffen it, and it has finer teeth than the large crosscut saw, usually 14 to 16 teeth per inch. The back saw, when properly sharpened and set, can make a smoother and more accurate cut than is possible with the large crosscut saw. It cannot be used for cuts deeper than the width from the teeth to the beam on the back because the thickness of the beam is considerably greater than that of the blade.

Another saw, sometimes called a dovetail saw, is similar to but smaller than the back saw and has considerably finer teeth (approximately 26 per inch). This saw is desirable for making very fine cuts and is generally preferred over the standard back saw for cutting dovetails.

With a good back saw, it is possible to cut a piece of

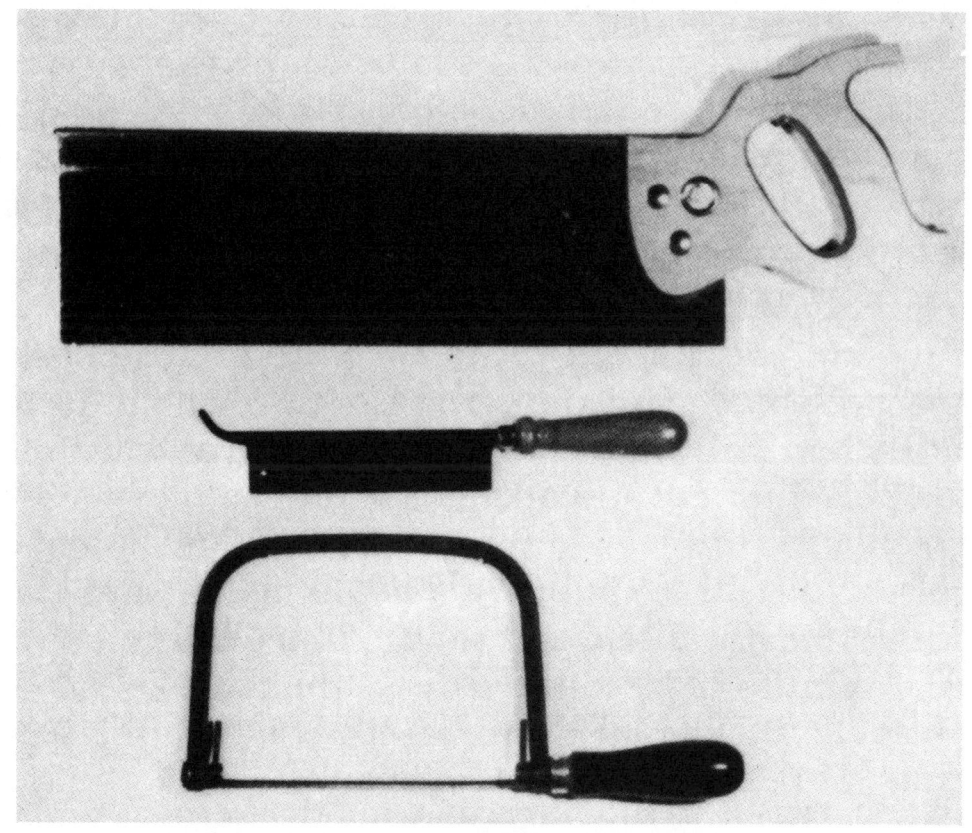

Fig. **2.1   A back saw, a dovetail cutting saw and a scroll saw.**

lumber very accurately to length. If the mark is made with a knife instead of a pencil, a good back saw can be controlled so that the teeth next to the mark will cut precisely on the mark (see Figs. 1.5 and 1.7). If the end of the piece being cut is to be a finished surface, the cut should be made at least 1/64" beyond the mark to permit the end to be planed or filed down to the mark. As the saw cuts, it tends to tear the wood fibers on the side of the kerf so some wood needs to be removed to eliminate these damaged fibers. The amount of wood which needs to be planed or filed away depends on the condition of the saw and the nature of the wood being cut.

The saws described above can be used only for straight cuts. When curves are to be cut by hand, a scroll saw should be used. The scroll saw has a D-shaped frame and blades having teeth of various degrees of fineness can be mounted in the frame. Tension can be applied to the blade by turning the handle. The types of saws previously described have their teeth pointing so that the saw cuts when it is pushed forward. The blade should be mounted in the scroll saw so that the teeth point back toward the handle so that they cut when the saw is pulled toward the operator. This is done because the D-frame is flexible in order to provide the tension in the blade and, if the cutting is done on the forward motion, the frame will flex and cause the tension to be relaxed from the blade during the cutting stroke. Scroll saw blades are quite inexpensive and should be replaced when they are damaged or become dull.

When curves are being cut with the scroll saw, the blade should be kept perpendicular to the lumber being cut. The blade clamps on the saw can be rotated so that the D-frame will not interfere with the piece of lumber being cut and, if a cut-out which does not open on the edge of the piece is being made, you can bore a hole near the line and assemble the blade in the saw after it has been placed through the hole.

The scroll saw blades do not cut as cleanly as a well-sharpened back saw and it is difficult to control it as accurately, so all cuts made with it should allow at least 1/32" from the mark to be removed with a file or a rasp.

It is possible to purchase a miter box consisting of a large back saw which operates in guides supported in a frame. The guides can be set so that the miter can be sawed at any desired angle. The piece to be cut can be clamped in place in the box with a C-clamp while it is being cut.

A good miter box provides a very precise means of cutting a miter. But it represents a large investment and an amateur who has a power saw which is capable of cutting accurate miters will be reluctant to invest in a good miter box. However, it is difficult to cut miters on a power saw as accurately as they can be cut with a miter box.

2.2.1 Sharpening and Setting Saws

In order for a saw to do its job properly, the teeth should be sharp and properly set. It is very important, in using hand saws, to avoid striking metal with the teeth. Avoid sawing lumber which may contain nails and do not use a metal support for lumber being sawed unless all metal parts of the support are at a safe distance from the saw cut. If much hand sawing is to be done, a pair of sawhorses should be constructed.

The first step in sharpening a hand saw is jointing. Jointing is the process of filing the ends of the teeth to

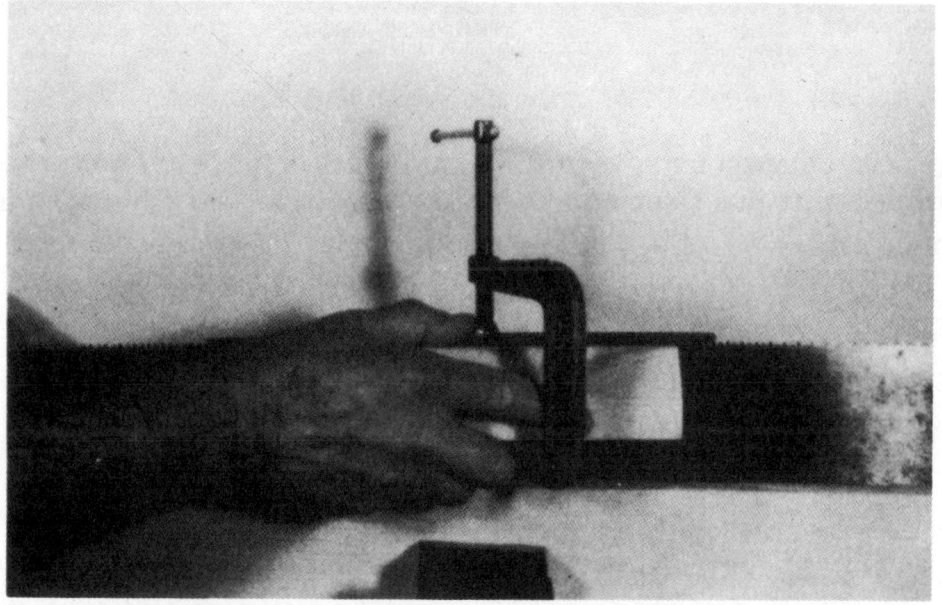

Fig. **2.2 Arrangement for jointing a hand saw.**

bring them all to the same level. The saw is jointed with a mill file held perpendicular to the saw blade by means of a wood block and a clamp, as indicated in Fig. 2.2. This procedure serves a two fold purpose. First, it reduces all of the teeth to the same level. Second, the ends of the teeth are filed to a flat so that when the cutting edge is filed it can be filed so that the flat surface at the top is just reduced to a point. The saw blade should be clamped between two narrow boards in a vise to hold it straight. Figure 2.3 shows how the teeth appear after jointing and filing.

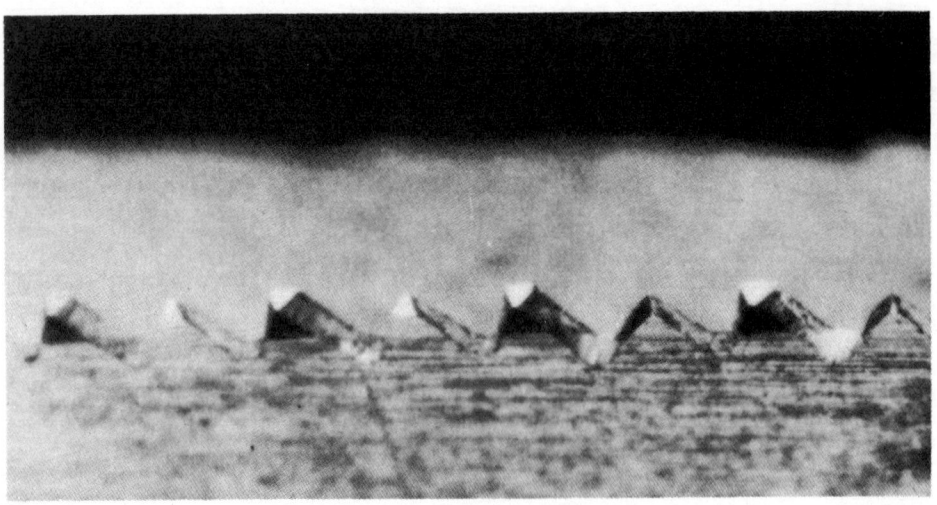

Fig. **2.3   Crosscut saw teeth magnified. The teeth have been jointed and the first and third on the right have been sharpened.**

The teeth are filed with a small triangular file. On a crosscut saw, the file should be stroked at an angle of about 65° relative to the tail end of the saw with the handle of the file down at about 10°. Take firm strokes and raise the file from the tooth for the return stroke. The flat surface produced by jointing is the reference level and the tooth should be filed until the flat just dis-

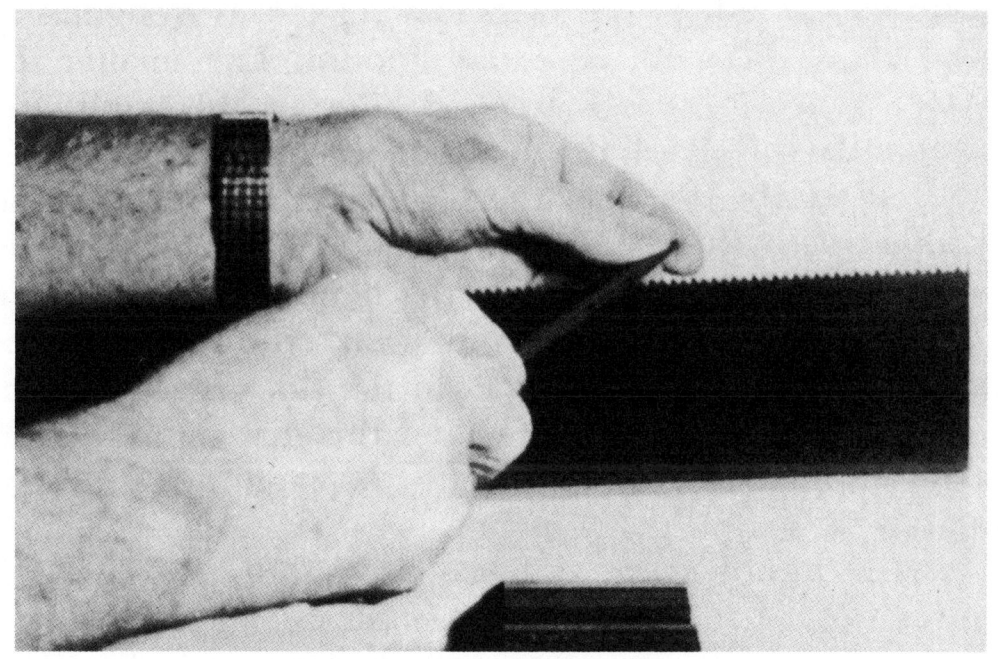

Fig. **2.4   Filing crosscut saw teeth.**

appears to be replaced by a point. Further filing will reduce the length of the tooth below the reference. The
bottom of the file will increase the depth of the gullet
while the tooth is being filed. Figure 2.4 shows the file
being used to file the teeth set to the side of the blade
toward the operator. In the same manner, file alternate
teeth over the entire length of the blade. When this is
completed, reverse the blade in the clamp and file those
teeth set on the other side.

A rip saw is handled in the same manner except that
the file is stroked nearly straight across with the file
handle about two degrees down.

After the teeth on the saw have been sharpened, it is
necessary to set them. This is done by means of a special
tool called a saw set. The teeth are set by bending alternate teeth out of the plane of the saw blade on one side
and bending the other teeth out of the plane of the

blade on the other side. The saw set tool is designed to bend all of the teeth an equal amount. The smaller the point count of the saw, (i.e., the larger the teeth) the greater the offset of the point of each tooth. To set a tooth, place the tool over the saw blade, which should be clamped with the teeth pointing up. The anvil on the saw set is on the side opposite the handles and therefore only the teeth which are bent away from the handles should be set. There is a dial on the saw set which controls the position of the anvil and this dial should be set at the number corresponding to the points per inch of the saw.

As the handles are squeezed together a small arm moves toward the anvil and the set should be oriented so that this arm strikes the tooth symmetrically. The handles are then firmly compressed to bend the tooth firmly against the anvil. Figure 2.5 shows the saw set in use. Since the saw set bends the teeth away from the operator, alternate teeth will be set with the saw positioned as in Fig. 2.5 and it will be necessary to turn the saw around in order to set the other teeth.

Fig. **2.5   A saw set in use.**

2.3 PLANES

There are three types of planes used for planing the lumber parallel to the grain. These planes are similar in construction and vary only in their length and the width of the cutting iron.

The smooth plane is about 9" long and has a 2" cutter. It is used for planing small pieces and for pieces with uneven surfaces. The jack plane is a general purpose plane which is about 14" long with a 2" cutter. It is the plane which should be selected if you will have only one plane in your shop. The fore plane is about 18" long with a 2 3/8" cutter. This plane is superior to the jack plane for planing large surfaces and the edges of long boards.

Figure 2.6 shows a plane iron assembly. The chip

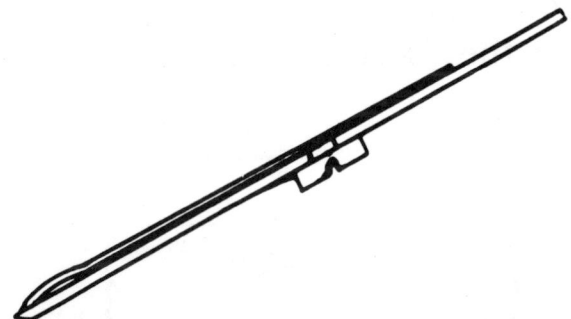

Fig. **2.6   A plane iron assembly.**

breaker or plane iron cap is clamped to the cutter blade with a large screw. The chip breaker should contact the cutter at least 1/32" from the cutting edge but not more than 1/16" from the edge, and the bottom edge of the chip breaker should be parallel with the cutting edge. The screw should clamp the chip breaker firmly against the cutter iron. When adjusting the position of the chip breaker, be very careful not to allow it to slide over the cutting edge of the iron as it will spoil the cutting edge.

The plane iron assembly is placed in the plane with the chip breaker on top and its position is adjusted until a small rectangular dog in the plane engages a rectangular opening in the chip breaker. This dog is one end of a lever which is controlled by the adjusting nut which is below and back of the cutting iron. This adjusting nut controls the projection of the blade below the bottom of the plane. A lever on top of the handle, just below the iron, controls the tilt of the blade in the slot in the bottom of the plane. The tilt of the blade should be adjusted so that its cutting edge is parallel to the edges of the slot so that the cutter will make the same depth of cut over its entire width. The depth of cut should be adjusted to produce the desired thickness of shaving.

Before starting to plane a piece of lumber, particularly if it is hardwood, examine the orientation of the grain. If the grain of the piece is as indicated in Fig. 2.7, the

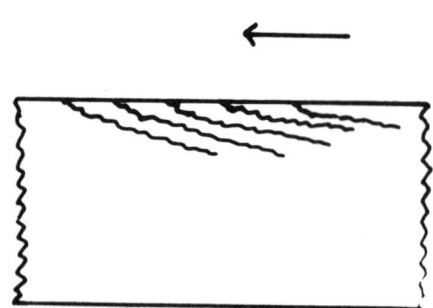

Fig. 2.7 **Illustrating the direction of planing relative to the grain in the lumber. The plane should be pushed in the direction of the arrow.**

plane should be pushed in the direction of the arrow to avoid chipping. If the grain is parallel to the surface, the piece can be planed in either direction.

Normally, the lumber is planed parallel to the grain. Large panels or table tops, formed by gluing several pieces together, may be uneven due to slight warp or twist in the individual boards. Such panels or tops should be planed across the grain. The long flat bottom on the plane will bridge the low places and the plane will remove wood from the high places. The surface left by the plane will not be very smooth but it can be smoothed after it is made flat by either planing with the grain or by sanding.

When a series of boards are to be glued together to form a panel or table top, the edges can be formed with a plane. If the panel is not over three feet long the jack plane can be used, but for longer panels the long fore plane should be used. The edges of two adjacent pieces which are to be glued together should be planed simultaneously while they are clamped in a vise as indicated in Fig. 2.8. By arranging the pieces as indicated in Fig. 2.8, the panel will be flat when the edges are clamped together, even if the plane produces a surface with a slight angle, because the angle error on one piece will compensate the angle error on the other.

After the surfaces have been planed, they should be matched together as in Fig. 2.9. The ends of the joint, *a* and *b*, should make contact while there is a very slight separation at *c* which tapers to zero as *a* and *b* are ap-

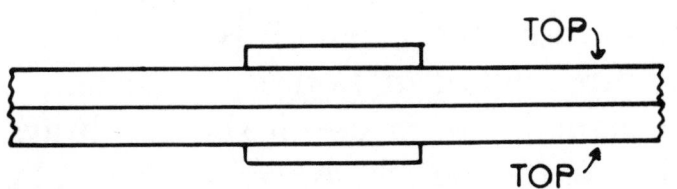

Fig. **2.8 Arrangement of pieces of lumber for planing to prepare them for an edge-glued joint.**

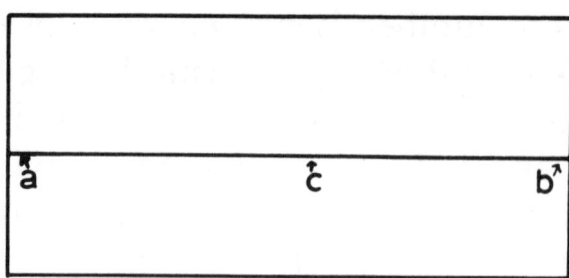

Fig. **2.9 Matching pieces of lumber for an edge-joined glue joint. They should make contact at** *a* **and** *b* **and be slightly separated at** *c*.

proached. This configuration can be achieved by pressing down on the plane slightly harder in the midrange of the stroke than at the two ends.

The planes described above are primarily suited for planing the surfaces or edges of the lumber. They are not as well suited for planing the ends, where the cut is made across the grain. For this operation a block plane should be used. The block plane differs from the planes described above in the following respects: (1) The plane is shorter and narrower. (2) The cutting iron does not have a chip breaker. (3) The blade makes a smaller angle with the bottom of the plane. (4) The ground bevel is on top of the cutter iron rather than on the bottom.

The adjustments of the block plane are similar to those of the other planes. When the end of a piece of lumber is planed with a block plane, care must be exercised to prevent splitting off a portion of the edge of the piece. This can be prevented by cutting a small bevel on the corner as indicated in Fig. 2.10, where the arrow indicates the direction of motion of the plane.

When a plane is not in use, it should always be placed on its side on the bench. *Never* set the plane on the bench on its bottom, because it will rest on the sharp

Fig. **2.10** **Use of a bevel to prevent splitting of the edge when planing the end of a piece of lumber with a block plane.**

edge of the cutter iron, which can injure the cutting edge. If you will cultivate the habit of always placing the plane on its side, it will soon become automatic to do so.

2.3.1 Sharpening Plane Cutter Irons

Figure 2.11 shows the ground edge of a plane cutter

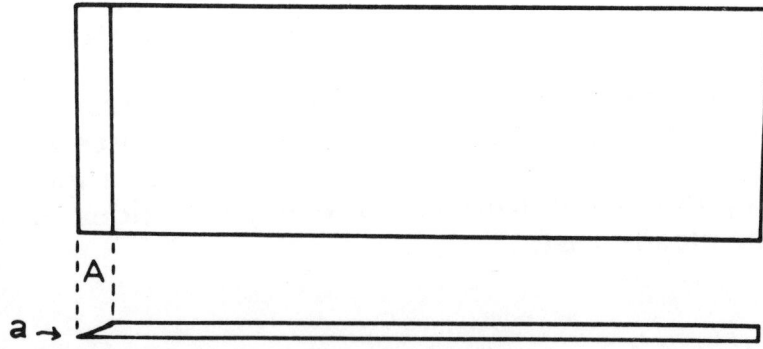

Fig. **2.11** **Ground edge of a plane iron.** *A* **is the width of the ground surface and** *a* **is the cutting edge.**

iron. The iron is ground on a Carborundum wheel so that the ground surface has the same radius as that of the grinding wheel. The length *A* of the ground portion should be about 2-1/3 times the blade thickness.

It is important to grind the iron so that the cutting edge is straight and perpendicular to the side of the iron. After it is ground it will need to be honed. It can be honed several times before it will be necessary to grind it again. Of course, if the edge is nicked, it will be necessary to regrind it.

Figure 2.12 shows the arrangement for accurately grinding a plane iron. The plane iron is shown at *A*. *B* is

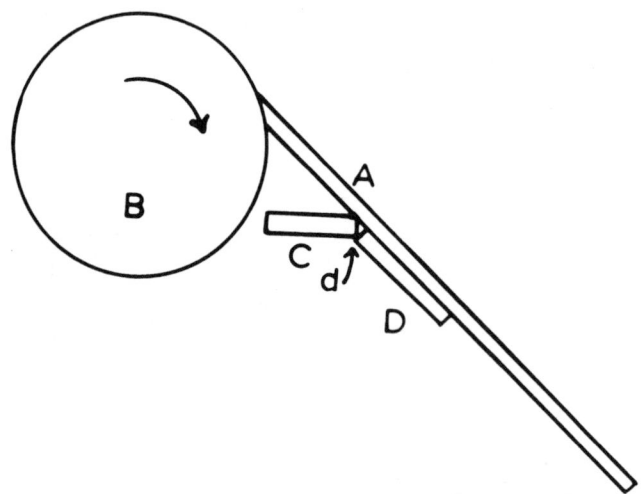

Fig. **2.12 Arrangement for grinding a plane iron.** *A* **is the plane iron,** *B* **is the grinding wheel which rotates in the direction of the arrow,** *C* **is the steady rest,** *D* **is a guide plate attached to the plane iron and the edge** *d* **slides along the edge of the steady rest.**

the grinding wheel which rotates in the direction indicated by the arrow. The steady rest on the grinder is indicated at *C*. The front edge of the steady rest must be smooth and it must be parallel to the axis of the grinding wheel. The plate *D* should have its edge *d* smooth and straight and it should be clamped to the iron so that the edge *d* is perpendicular to the side of the iron. It is often possible to use the chip breaker for the plate *D* by placing it on the back and turning it perpendicular to the iron. When the iron from a block plane is ground, the metal plate D can be a rectangular plate secured to the iron by means of a small C-clamp. The clamping screw is tightened when the position of the chip breaker is such that the iron makes contact with the wheel so that the ground surface will be like that shown in Fig. 2.11.

After the plate D is clamped to the iron, make a light pass over the wheel and check to determine if it starts to grind the bevel correctly. It may take two or three trials to find the proper adjustment.

You should have a dish of water handy and then *put on your safety glasses* and start the grinder. The blade should be held lightly against the wheel and symmetrically moved from side to side so that metal will be removed uniformly along the entire ground portion. You should frequently dip the iron in the water to cool it. If you try to grind the iron too fast it will burn. This is indicated by the metal at the edge turning blue or black. Burning is most apt to occur when the grinding has proceeded to the point where the hollow ground portion just approaches the cutting edge of the iron.

After the plane iron has been ground, it will be necessary to hone it. It should be held on the honing stone so that the cutting edge in Fig. 2.11 is in contact with the stone and the back edge of the hollow-ground portion is just above the surface of the stone. A fine Carborundum stone with a flat surface should be used and oil should be applied to the surface. The usual method of honing is to impart a circular motion to the iron while it is in contact with the stone. The honing should be continued until it is possible to see a honed portion at the front of the hollow-ground portion of the iron. If you touch the top of the edge with your finger you can feel where the metal at the edge has been "turned over" in the process of grinding and honing. This turned-over metal can be removed by turning the iron over, holding it nearly flat against the stone and honing lightly until it is no longer possible to feel any roughness on the top side of the edge.

If you have a fine arkansas stone, a finer cutting edge can be obtained by finish honing on the arkansas stone.

It is important to maintain a constant angle between the iron and the stone while honing. A beginner may have difficulty achieving this. It is possible to purchase a holder for the plane iron like that shown in Fig. 2.13.

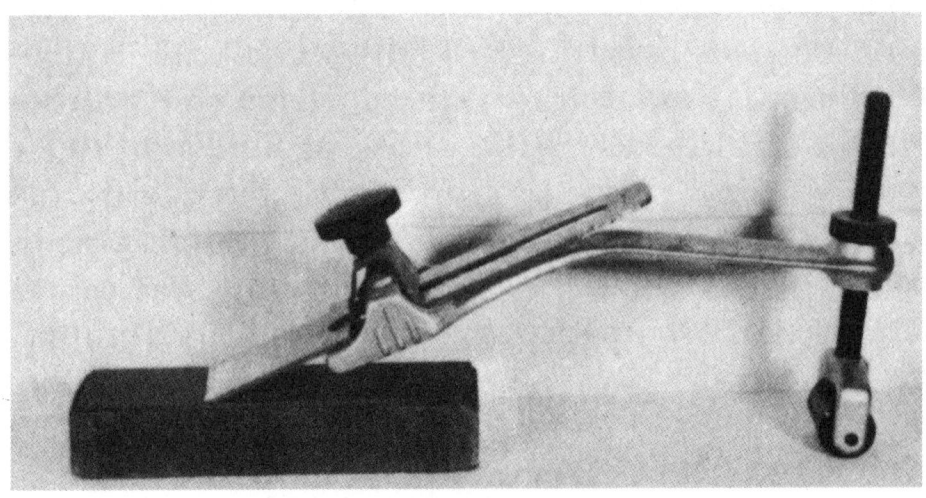

Fig. **2.13   A holder for honing a plane iron.**

The stone and the wheel should be on a flat surface and the iron is moved back and forth along the length of the stone. I like to use the flat iron surface of my table saw for this purpose.

2.4   HAND BORING TOOLS

The brace and auger bit is the most precise tool for

boring a hole of fixed diameter about a fixed center. The brace consists of a head with a quill bearing which can be held by the operator, a sweep handle for turning and a chuck for holding the auger bits. The brace may or may not have a ratchet just above the chuck. A ratchet makes it possible to bore holes in cramped quarters where there is not space available to swing the sweep handle completely around the center of the hole. There are many instances where the ratchet is necessary to make it possible to bore the hole, so when purchasing a brace, it is poor economy to purchase one without a ratchet. The ratchet control can be set so that the ratchet is inoperative or it can be set so that it is operative either for boring the hole or backing the bit out of the hole.

The auger bits have a square-tapered tang on one end for clamping the bit in the chuck. Below the tang is a straight shank which leads to the twist which has the same diameter as the hole being bored. The twist serves to carry the shavings out of the hole. At the bottom of the twist are the lips, which cut the wood at the bottom of the hole. At the outside edge of each lip is a spur or circumferential cutter which cuts the wood around the circumference of the hole. At the center of the bottom of the bit is a feed screw which serves to feed the bit into the hole. The feed screw tapers to a point at the end and serves to accurately feed the bit to bore a hole centered where the feed screw was started. The feed screw is especially important for boring in a nonhomogeneous material such as wood.

Auger bits can be purchased in sizes from 3/16" to 2", but sizes larger than 1" are seldom used because expansion bits are available for boring holes from 1" to about 3" in diameter. The expansion bit is similar to the standard auger bit. It has a feed screw but has only one lip and spur. The lip and spur are on a single piece which is adjustable to vary the diameter of the hole. When the lip is moved out to the proper radius, it is clamped by means of a screw at the bottom of the shank. In order to cover the entire range of hole diameters, two sizes of lips are available. The long lip is used for boring the larger sized holes.

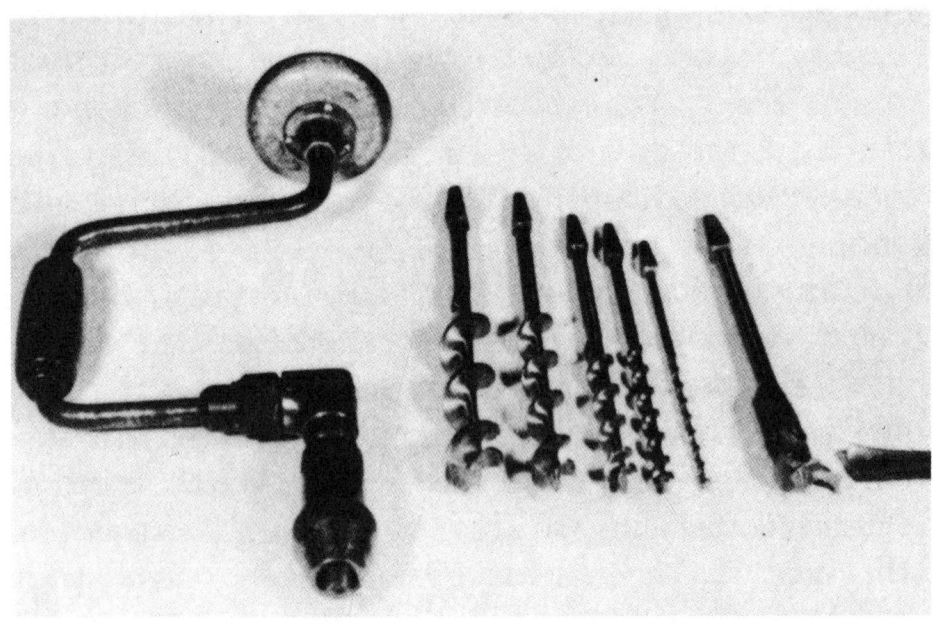

Fig. **2.14** **A brace, auger bits and an expansion bit.**

Figure 2.14 shows a brace, a collection of auger bits and an expansion bit. The additional lip and spur is shown beside the expansion bit.

Standard auger bits are available in steps of 1/16". A number is stamped on one side of the tang to indicate the number of sixteenths of an inch in the diameter of the bit. For example, a bit marked number six bores a hole 6/16 or 3/8" in diameter.

It is possible to purchase a gauge to be attached to the shank of a bit which can be set for a specified hole depth. Such a gauge is not really necessary. I find that it is just as convenient to attach a piece of masking tape around the bit at the point where the boring should stop.

When a hole is bored through a piece of lumber, there is a tendency for the wood to split away from the hole on the back side when the bit comes through. This can be avoided if the hole is bored until the feed screw just comes through on the back side. The piece of lumber can then be turned around and the feed screw started in the hole where it came through. The remainder of the hole can then be cleanly bored from the back side.

When boring a hole in the edge of a piece of lumber near the end, there is danger of the piece splitting due to the sidewise pressure of the bit. This can be prevented by clamping across the piece in the region of the hole with a C-clamp.

It is often important to bore a hole perpendicular to the surface in which it is started. It is not easy to judge this, but it is helpful to place the lumber so that the auger bit and brace will be vertical. A try square can then be placed beside the bit and the edge of the try square will serve as a reference.

It is sometimes necessary to bore a hole at an angle. To do this, set the sliding "T" bevel at the correct angle with the protractor. Start the feed screw with the auger bit held vertically and then tilt it to the proper angle, using the sliding "T" bevel as a reference.

When a hole is to be bored, the location of the hole is determined and marked with a scratch awl or center punch. When a dowel pin joint is to be prepared, it is necessary that the holes in the two pieces match to high precision. This can be accomplished by boring the holes in one of the pieces. They can be brought together in the correct orientation with a dowel center, like that shown in Fig. 2.15, placed in one of the holes. The

 *Fig.* **2.15   A dowel center.**

pieces are then pressed together so that the point on the dowel center will mark the other piece. The hole marked by the dowel center is then bored and the dowel center is placed in the other hole and the joint is assembled with a dowel pin in place in the first matching holes. This will accurately mark the position of the second hole which can then be bored. A dowel pin joint, using two dowel pins, is often used instead of a mortise and tenon joint. It is very important, when boring the holes for such a joint, to bore the holes perpendicular to the surface.

Usually edge-glued joints in panels or table tops are reinforced with dowel pins placed about 8" or 10" apart along the length of the joint. Accuracy of placement of the dowel pin holes is important to insure that the joint will go together when it is assembled with the dowel pins in place. It is also important that the spacing of the pins from the top side of the panel be the same in order to insure that the pieces forming the panel will all be in the same plane. To insure this, the two pieces to be glued together are clamped in a vise and the locations of all of the dowel pins are marked with a try square and a sharp pointed pencil as indicated in Fig. 2.16. A dowling jig, like that shown in Fig. 2.17, is clamped across the two boards with its index on one of the lines shown in Fig.

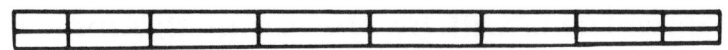

Fig. **2.16** **A pair of pieces of lumber to be edge-glued, marked for boring the dowel pin holes.**

2.16. The bit guide, corresponding to the bit size to be used, is placed over the line and adjusted to the approximate center of the board and the hole is bored. The doweling jig is then removed and turned around and clamped in position with the index at the other end of the line and the second hole is bored. This insures that the holes will be centered on the lines and the two holes

Fig. **2.17** **A dowling jig in use.**

for each dowel pin will be the same distance from the top of the panel.

If you do not have a doweling jig, all of the holes can be bored in one piece and a dowel center can be placed in the hole at one end. The boards can be pressed together with the top surfaces in the same plane to mark the first matching hole in the second board. Bore this hole and insert a dowel pin. Now put the dowel center in the hole at the other end of the first board and with the dowel pin in the first pair of holes, mark the center of the matching hole in the second board. After placing a dowel pin in this pair of holes it is possible to mark the centers of the remaining holes in the second board with the dowel center. The doweling jig helps to make it certain that the holes will be bored perpendicular to the

Fig. **2.18   A hand drill.**

surface. If the holes are marked with the dowel center, care must be exercised to be certain that the holes will be bored perpendicular to the surface. You should use your try square to insure this.

When holes smaller than 3/16" are to be bored by hand, the hand drill shown in Fig. 2.18 can be used. Twist drills similar to those normally used for drilling metal are usually used in the hand drill, although it is possible to obtain specially shaped twist drills which have a tip like that shown in Fig. 2.19. The point at the center of the tip is a guide point which permits the drill

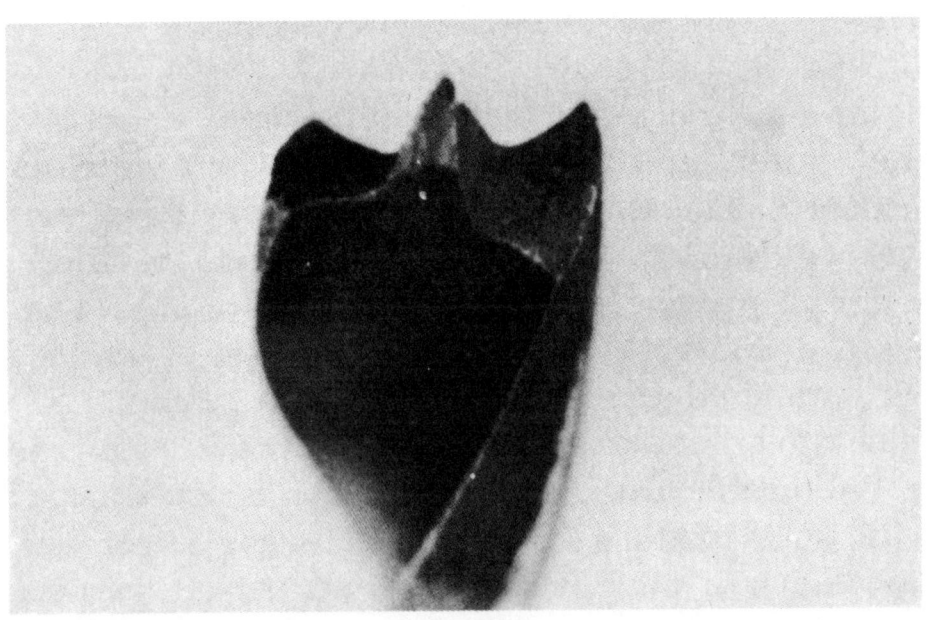

Fig. **2.19   A special wood-boring twist drill.**

to drill holes in wood with greater accuracy of center than is possible with a standard twist drill. We will cover twist drills in more detail when we discuss boring with power tools.

2.4.1   Sharpening Auger Bits

Figure 2.20 shows the cutting end of an auger bit.

Fig. **2.20   The cutting end of an auger bit.** *A* **is the spur,** *B* **is the lip and** *C* **is the feed screw.**

The lip or chisel cutter at *B* and the spur or circumferential cutter at *A* are the portions which need to be sharpened. The lip can be sharpened by filing on the bevel with a narrow file. The lip should be filed just enough to provide a sharp edge and care should be exercised to file both lips the same amount to insure that their cutting edges are both in the same plane.

The spur should be filed on the *inside* with a small file. Do not file away more metal than is necessary. I like to use small pattern files for sharpening auger bits because they are small, which is an advantage, particularly when sharpening small bits, and they are very fine, which enables them to produce a fine edge.

2.5   CHISELS AND WOODCARVING TOOLS

Chisels have blades with straight cutting edges. The chisels commonly used have blades with widths of cutting edges from 1/4" to 1". The cutting edges are ground with a bevel similar to that on a plane iron and the bevel is usually 3/16" to 1/4" wide.

Figure 2.21 shows an ordinary set of chisels of 1/4", 1/2", 3/4" and 1"

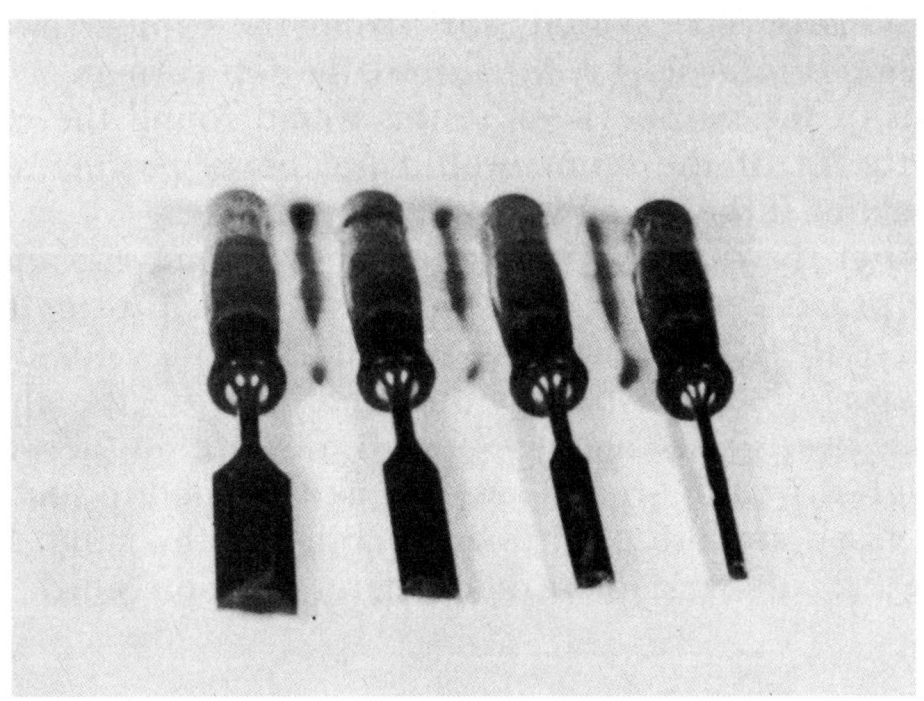

Fig. **2.21 An ordinary set of chisels.**

When mortises are being cut by hand, they are layed out as indicated in Fig. 2.22 and a series of holes are bored with a diameter equal to the width of the mortise and to a depth equal to the depth of the mortise. The excess wood is then removed with chisels. The cutting edge of the chisel is placed on the line bounding the edge of the mortise with the bevel side toward the

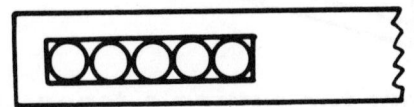

Fig. **2.22 Layout for cutting a mortise with the holes already bored.**

center of the mortise, and it is driven straight down with a mallet. A wide chisel is used on the long sides and when they have been cut down about 1/8" a narrow chisel is used at the ends. The narrow chisel is then used to prize the waste wood out to the depth of the cuts with the chisels. You should work from the center toward each end to avoid prying against the top corners at the ends of the mortise because this would round the edge so the fit of the tenon would not be as perfect as it would be if the edges were square.

After the first layer of waste wood is removed, repeat the process until the waste wood is removed to the bottom of the bored holes. Avoid trying to remove too deep a layer of waste wood at one time because the chisel blade exerts considerable outward pressure on the wood as it is driven down and it can cause the wood to split.

Chisels are also used for cutting tenons by hand. Figure 2.23 shows a piece marked for a tenon which has

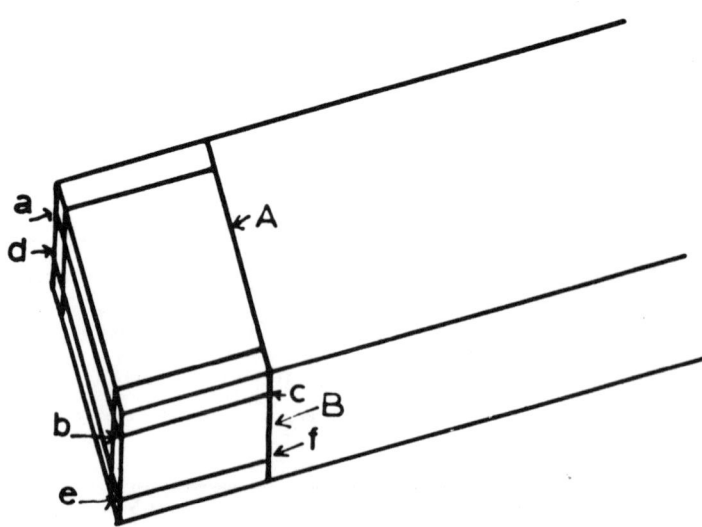

Fig. **2.23   A layout for cutting a tenon.** *A* **and** *B* **are the saw cuts.** *abc* **and** *def* **define the planes to which waste wood must be removed.**

the saw cuts already made at *A* and *B*. The wood above the plane bounded by *a, b* and *c* and that below the plane bounded by *d, e* and *f* can be removed with a wide chisel. In the same manner, the waste wood on the narrow sides of the tenon can be removed. The waste wood should be removed in layers and the chisel should be gently driven with a mallet.

Chisels may also be used for the final shaping of curved pieces, as indicated in Fig. 2.24. The chisel should be driven in the direction indicated relative to the grain of the wood. If it is driven in the opposite di-

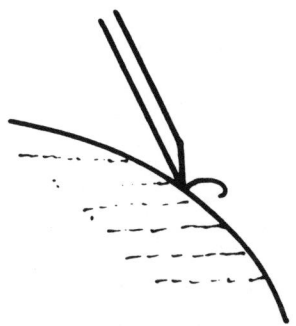

Fig. **2.24** **Shaping a curved piece of lumber with a chisel.**

rection the wood will tend to split along the grain. In an operation such as that indicated in Fig. 2.24, it is usually better to drive the chisel gently with a mallet. If the chisel is pushed, it is difficult to control it. *Never* push the chisel toward your other hand as you will almost certainly injure yourself. It is sometimes necessary to push a chisel. When this is done, if you grasp the handle of the chisel with your right hand you should grasp the blade with your left hand to guide it. It is important to keep the chisel well honed for such an operation.

There are some instances where some carving must be done on a piece of furniture. For this purpose, it is necessary to have some carving tools which may consist of curved gouges of different sizes, a V-gouge, knives and chippers. The gouges and knives are most important.

Those interested in woodcarving as a hobby should consult a book on woodcarving such as *Handbook of Woodcarving and Whittling* by Elsie V. Hanauer, published by A. S. Barnes and Company; *Whittling and Woodcarving* by E. J. Tangerman, published by Dover Publications; or *The Craft and Creation of Wood Sculpture* by Cecil C. Carstenson, published by Scribner's Sons.

When you wish to remove wood from a concave surface, a gouge should be used and a gouge with a radius of curvature less than that of the surface being formed should be selected. The gouges are used in the same manner as the chisels.

When wood is to be chipped out of small areas, the chipper or a knife should be used and a V-gouge should be used for cutting fine V-lines.

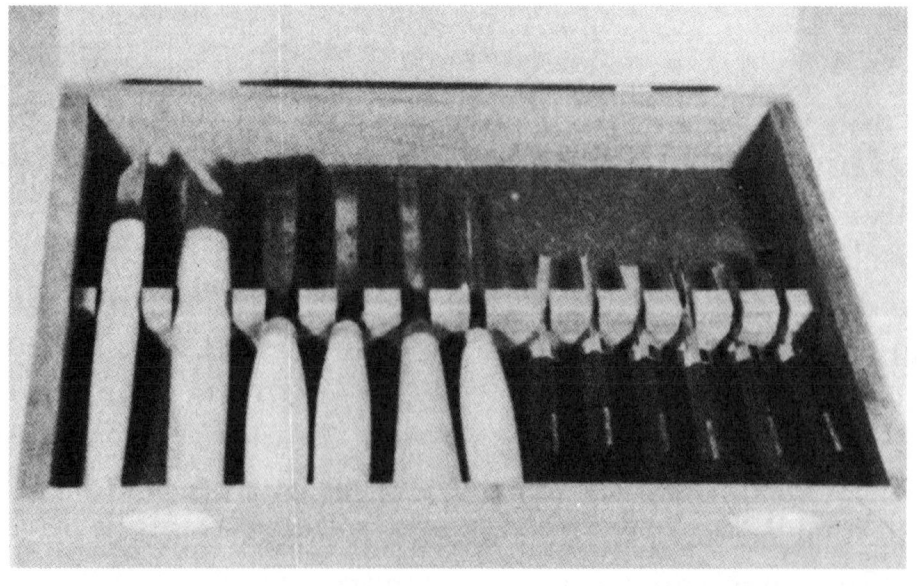

Fig. **2.25 A selection of woodcarving tools.**

Figure 2.25 shows a selection of woodcarving tools that is adequate for any carving that will be necessary in furniture construction.

2.5.1 *Sharpening Chisels and Woodcarving Tools*

Chisels are sharpened in the same manner as plane irons. When you have developed sufficient skill, it is possible to use your finger to replace the plate *D* in Fig. 2.12 when grinding a chisel, or you can clamp a plate to the top of the chisel blade to insure obtaining a straight cutting edge on the blade.

The chisel must be honed quite often when it is being used in order to maintain a keen cutting edge. A chisel is honed in the same manner as a plane iron.

A gouge can be ground on the side of the grinding wheel by resting the gouge on the steady rest *A* in Fig. 2.26 and rotating it as the bevel is being ground. The

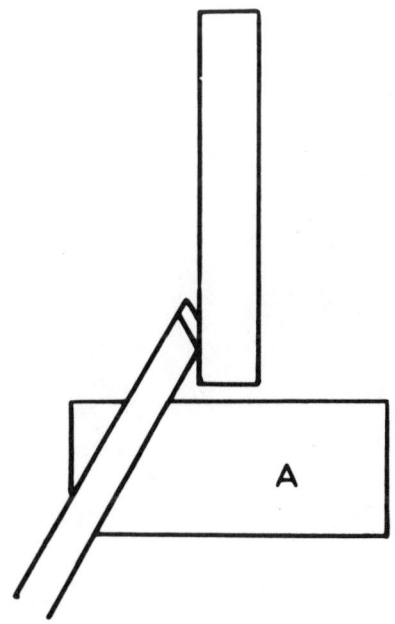

Fig. **2.26 An arrangement for grinding the bevel on a gouge.**

gouges are honed on a concavo-convex stone. The convex side of the gouge is honed on the concave side of the stone and the concave side of the gouge is honed on the convex side of the stone.

3

POWER SAWS

3.1 THE TABLE SAW

The most important power tool in the shop is a power saw. The table power saw is a highly versatile tool because of the accessories available for use with it. Its primary function, however, is for ripping and crosscutting lumber.

Figure 3.1 is a photograph of a table saw with the

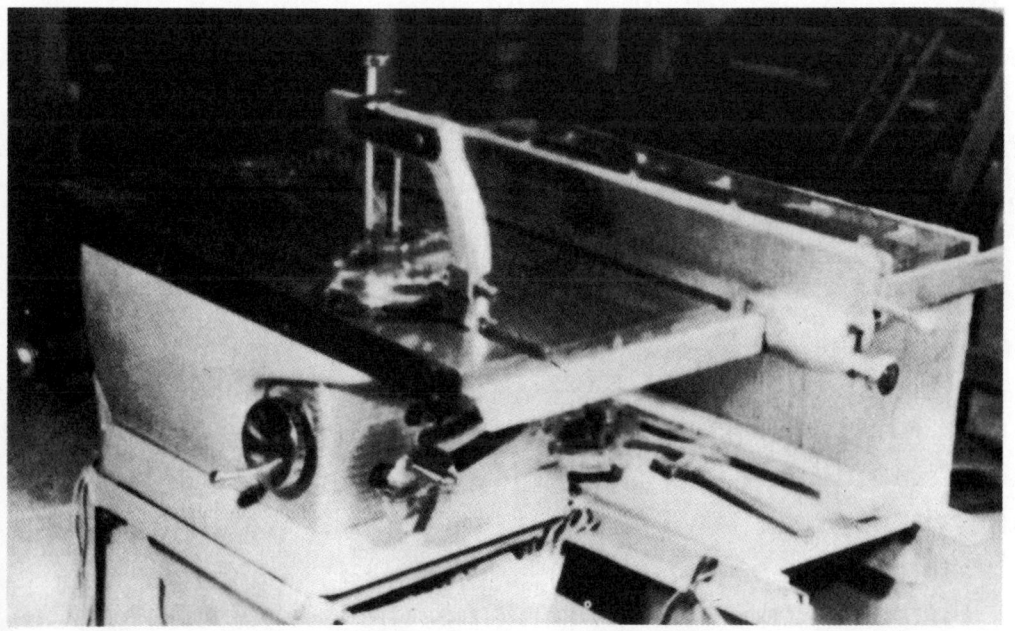

Fig. **3.1 A typical table saw with both the fence and the miter in place.**

fence and miter both in place on the table. The saw arbor runs in two bearings. It has a drive pulley at one end and the other end is threaded so that the saw blade can be mounted on it with a nut and washer. The arbor is mounted beneath the cast-iron table and may be fixed to the frame with provision for raising or lowering the table or the table may be fixed with provision for raising or lowering the arbor.

The motor for driving the saw is mounted on a plate which is attached to the frame carrying the arbor. The saw may be operated with a 1/2-horsepower motor but a 3/4-horsepower motor is more satisfactory. The size of the motor determines the maximum cutting speed. The motor cord should be a three-wire cord with the ground wire securely grounded on the motor. The plug and receptacle should be the three-wire type with the ground terminal properly grounded in the receptacle box. If you do not have three-wire receptacles, have your electrician install them at all of the outlets in your shop. A V-belt is used to drive the arbor shaft by the motor. The optimum speed for an 8" saw blade is about 3,400 rpm, while that for a 10" saw blade is about 3,100 rpm. To drive the arbor at the proper speed, the pulleys for the motor and arbor shaft should be so selected that the speed of the motor multiplied by the ratio of the pulley diameter on the motor shaft to the pulley diameter on the arbor shaft is equal to the desired speed of the saw blade.

The saw blade projects through an opening in the table called the throat. Plates with various types of slots to accommodate the various cutter blades are used to close the throat and provide a surface flush with the top of the table.

There is provision for tilting the blade through an

angle of 45° relative to the surface of the table. In saws that are arranged so that the table is raised and lowered relative to a fixed arbor, the tilt is provided by tilting the table, while in saws that have a fixed table, the tilt is produced by tilting the arbor. The former type is called a tilting-table saw while the latter type is called a tilting-arbor saw. The tilting arbor saw is generally more convenient to use, especially when cutting long pieces at an angle. When a saw and jointer are mounted in combination to be run by a single motor, a tilting table saw is used because in a tilting arbor saw it is necessary to tilt the motor along with the arbor.

The fence, on the right-hand side of the table in Fig. 3.1, is used to guide the piece of lumber when it is being ripped; the miter, on the left-hand side of the table, is used to hold a piece of lumber when it is being crosscut. Normally, when the fence is being used, the miter is removed from the table and vice versa.

It is possible to purchase a variety of types of saw blades. The blades most useful for amateurs are the combination blades which can be used for both ripping and crosscutting. Figure 3.2 shows a selection of combination blades for use on a table saw. The blade at A is a carbide-tipped blade. The tip of each tooth has a piece of carbide welded in place and ground to the proper shape. This is a very useful blade for all rough cutting and rarely needs resharpening. If the blade does need resharpening, it is necessary to send it back to the manufacturer. Do not try to sharpen a carbide-tipped blade yourself. The blade at B is a tapered blade. This blade tapers in thickness from the teeth toward the center but has a full thickness center where it is attached to the arbor. This blade does not require the teeth to be set and it produces a very smooth finish cut. It should be

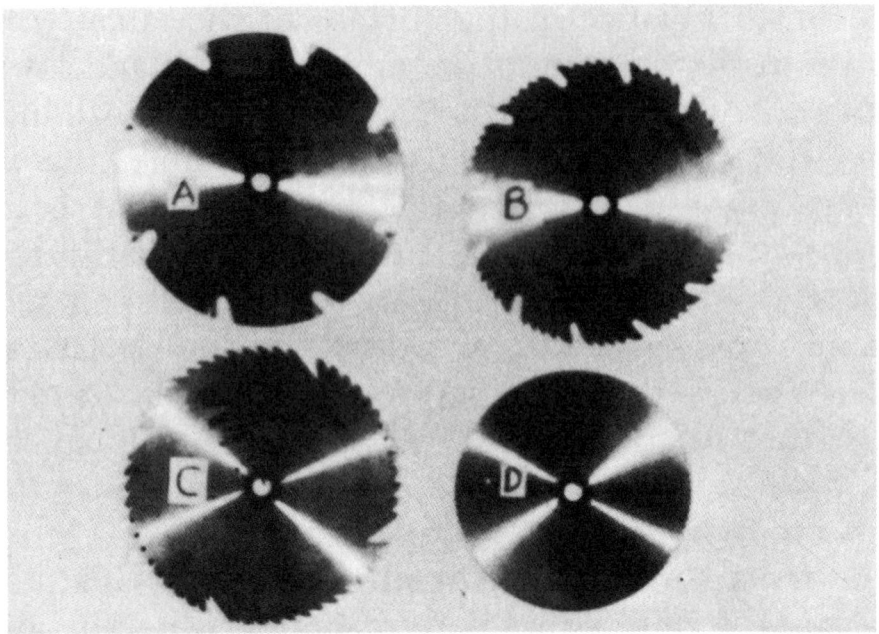

Fig. **3.2 Some typical circle saw blades.** *A* **is a blade with carbide-tipped teeth,** *B* **is a tapered blade,** *C* **is a tenon-cutting or miter-cutting blade and** *D* **is a plywood-cutting blade.**

used only for cuts where a very smooth cut is desired, since it is necessary to cut very slowly to avoid overheating the blade. The blade at *C* is a tenon-cutting blade. This blade is also useful for general cutting. It produces a smoother cut than the carbide-tipped blade if it is properly sharpened and set. It can be resharpened by the user. The blade at *D* has special fine teeth for cutting plywood. Blades *A* or *C* can be used for rough cutting plywood but the blade *D* will produce a smoother cut.

The arbor on a saw may have a diameter of 1/2", 5/8" or 3/4". Most blades are provided with "knockouts" in the center hole. A given blade may come with a 1/2" center hole but have a "knockout" to provide a 5/8" center hole. When purchasing a blade, be certain that it

can be fitted to your saw arbor. Blades are available in
6", 7", 8", 10" and 12" diameters. The size designation
of the saw indicates the largest diameter blade that can
be used in it. If you have a 10" saw, much of your cutting can be done with an 8" blade. It is usually desirable
to use the smaller blade if the depth of cut will permit it,
as the cut will be made closer to the center of the blade
so there will be less error due to deflection of the blade.

3.1.1   Ripping with the Table Saw

Figure 3.3 is a diagram of the top of the saw table.

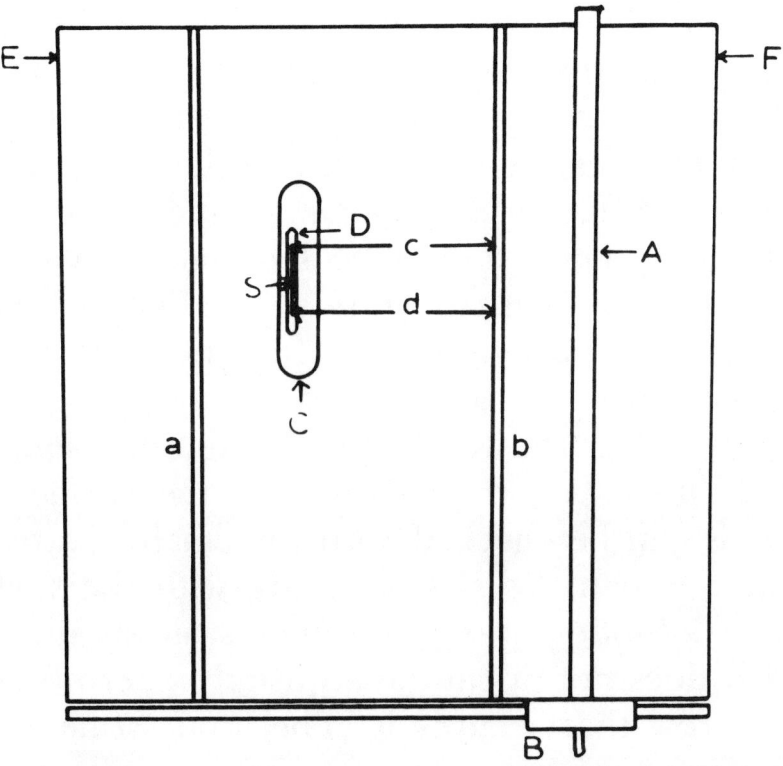

Fig. **3.3   Diagram of the top of a saw table.** *A* **is the fence,** *B*
is the clamp for clamping the fence, *C* **is the plate for clos-
ing the throat in the top of the table,** *D* **is the slot for the saw
blade,** *S* **is the saw blade and** *a* **and** *b* **are the grooves for the
miter gauge.**

The fence is indicated at *A*, and *B* is the clamp for clamping the fence in any position along the table. The plate which covers the throat is indicated at *C*, and *D* is the slot in the plate through which the saw blade *S* projects. The two dadoes in which the miter slides are indicated at *a* and *b*.

Before doing any ripping, it is necessary to properly align the saw table and the fence with the saw blade. The first step is to check the alignment of the table. To do this, raise the saw blade to its highest level and measure *c* and *d*. If these two measurements are not equal, loosen the bolts which hold the table to the frame and tap the table at point *E* or *F* until the measurements *c* and *d* are equal. When this is achieved, tighten the bolts, holding the table to the frame. Now move the fence to the edge of the dado *b* and adjust the fence in its clamp until it is accurately parallel to the edge of the dado. The screws which hold the fence relative to *B* will be different on different saws but it should not be difficult to find them. They should be loosened and the fence clamped so that it is accurately parallel to *b* and the screws can then be tightened. After the table and fence are adjusted, the saw can be used for ripping. To be certain that you cut the lumber with the edge square, the saw blade must be perpendicular to the surface of the table. This can be checked with the small square or the try square. When the saw is square with the surface of the table, the index on the tilt-angle scale should read at zero. If it does not, it can be adjusted to zero by loosening the screw that secures it. The angle scale will then read angles correctly.

If the blade on the saw needs to be changed, remove the plate *C* (Fig. 3.3) and loosen the nut on the arbor with a wrench. To hold the arbor while the nut is loosened, allow the shaft to turn until a piece of wood, such as a piece of dowel stock, can be placed across the opening in the table so that it engages a tooth on the saw. Remove the nut and washer and the saw blade. The desired saw blade should be placed on the arbor so that the teeth on the operator's side point down. Replace the washer and nut and tighten the nut with the wrench. You can hold the saw blade with your left hand while you tighten the nut with the wrench in your right hand. The nut tends to tighten as the saw is used. If the saw should be started while you are changing blades, you could be seriously injured, so you should always pull the plug on the motor cord from the receptacle while changing blades or making any adjustments where your hands come in contact with the blade.

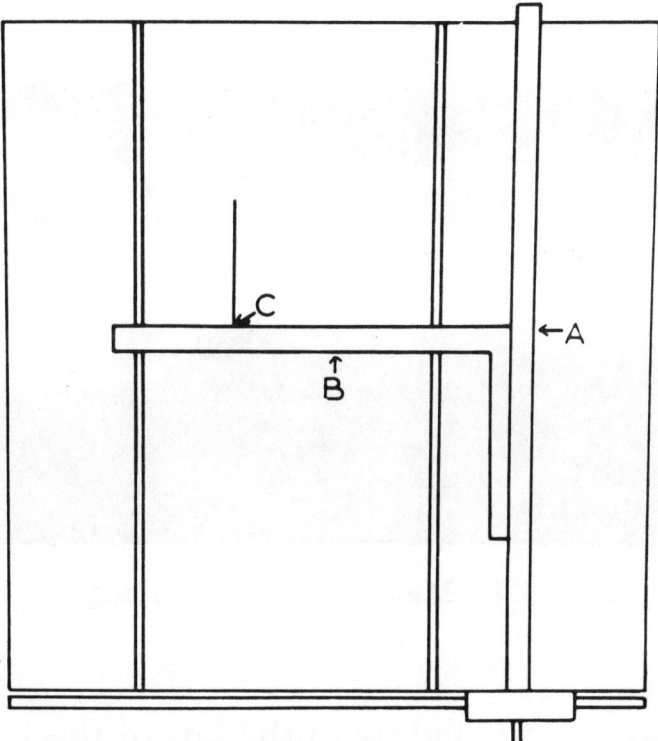

Fig. 3.4 Diagram showing the use of the large steel square for determining the width of a rip cut.

With the blade on the saw, you can set the fence for the desired width of cut with the large steel square as indicated in Fig. 3.4. A tooth, set toward the fence, should contact the scale on the square at the division indicating the width of the cut. If the saw has no set, the edge of any tooth can be used for the measurement. Clamp the fence and check the measurement again.

If the width between the saw blade and the fence is four inches or more, you can hold the portion of the piece of lumber between the saw and the fence with your right hand. To keep your hand away from the saw, hook your little finger over the fence. If the width of the cut is less than four inches, use a push stick in your right hand. Figure 3.5 shows the operation when a wide cut is being made and Fig. 3.6 shows the use of a push stick when making a narrow cut.

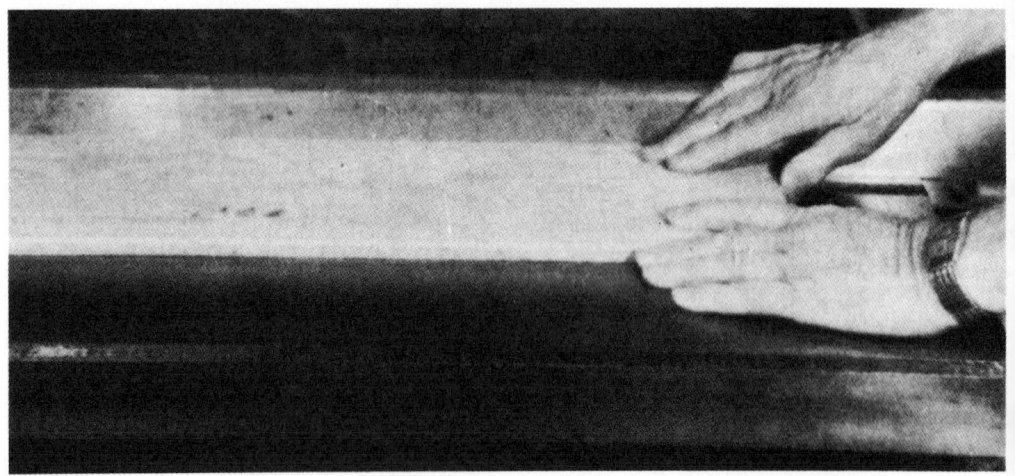

Fig. **3.5 Making a wide rip cut.**

You should stand slightly to the left of the saw and do not allow any onlookers to stand behind you for the saw may kick back the piece and cause an injury. When rip-

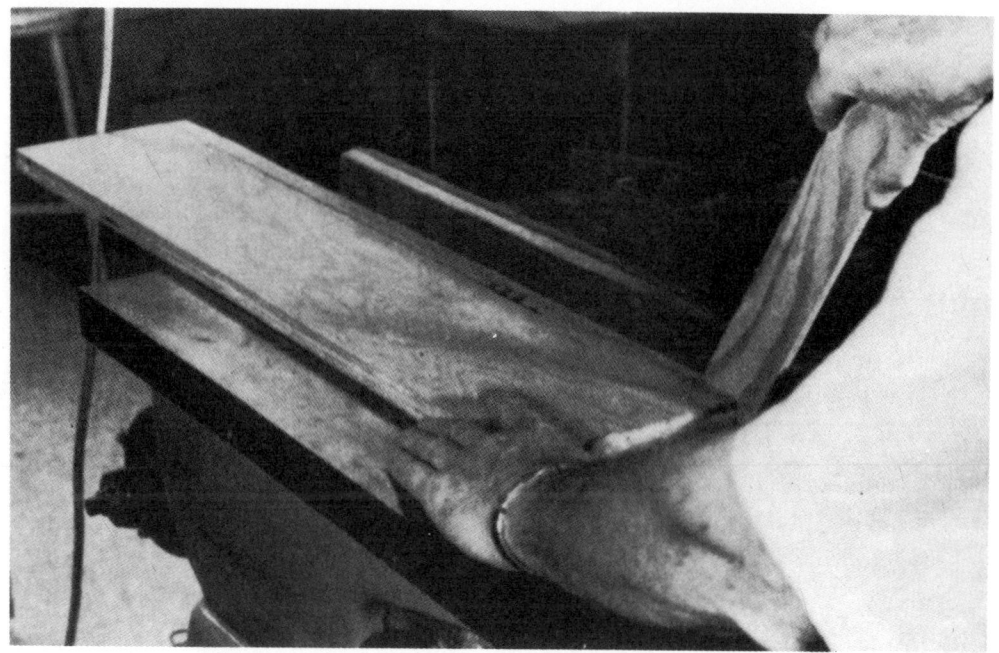

Fig. **3.6**   **Making a narrow rip cut using a push stick.**

ping short pieces, do not reach over the saw to retrieve them. Let them fall off the back of the table. The piece of lumber must be firmly held down on the table as the forward end approaches the far side of the saw blade. The friction of the saw in the cut tends to lift the forward end from the table and, if this is not countered, the saw will throw the piece back at the operator. This tendency is greater as the saw blade projects higher above the piece being cut. It is, therefore, especially important to be certain that the saw blade does not project more than 1/4" above the piece being cut, especially when short pieces are being ripped. When ripping long pieces, have someone steady the long end as it comes off from the saw table or you can construct a supporting stand like that shown in Fig. 3.7 for supporting the end of the piece of lumber as it comes off from the back of the saw table. The support *A* may have a roller mounted

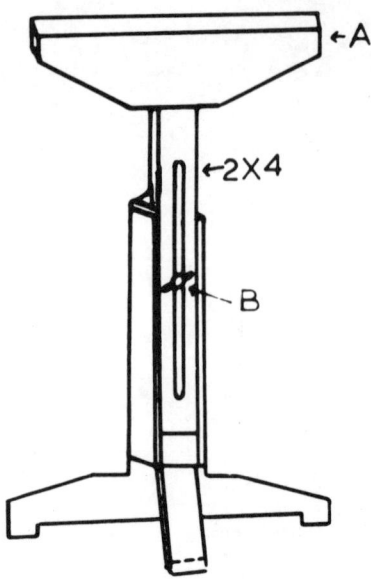

Fig. **3.7** **A stand for supporting the end of a long piece of lumber being ripped.**

in bearings at the ends or a row of ball-type casters spaced about 4" apart, with the balls facing up, for the lumber to ride on. The wing nut, with a flat washer, at B may be used to adjust the level of the support A.

3.1.2   The Miter Gauge

The miter gauge is an important tool for use with the table saw. It consists of a rectangular bar, which can slide in either dado groove a or b in Fig. 3.3, with a head which can be rotated through + or − 60° and that can be clamped to the bar at any angle. There is an angle scale with an index on the miter head but it is not possible to read the scale to very high precision. Most miter gauges have stops at 90°, + 45° and − 45°. The stops are actuated by means of a unit on the bar or handle which can be pushed into position to stop a screw on the rotating head at any one of the three positions. The arrangement varies somewhat with different miter gauges but the parts are relatively easy to identify. All good miter gauges have a clamp which can be used to hold the piece of lumber securely on the miter gauge. The clamp is not necessary when crosscutting at 90° but is desirable even with this operation. It is not possible to hold the piece securely enough with your fingers when cutting lumber at an angle, because the force exerted by the saw blade will cause the lumber to creep.

You should first check the adjustment of the 90° stop. To do this, set the miter at 90° by turning it against the stop and securing it in this position. Now take a piece of lumber 3" or 4" wide and 18" to 2' long, with one straight edge. With the straight edge against the face of the miter, saw off about 1/2" from the end of the piece. Check the squareness of the cut with the try square, being certain that the reference edge used is the same edge that was held on the face of the miter. The squareness check can be made very precisely if the square is held in position and you look toward a light with the piece of lumber and the square held so that any failure of the square blade to contact the end of the piece will show as a wedge of light. If the blade does not contact the end of the piece over the entire width, make a slight adjustment of the 90° stop screw. This is a cut and try adjustment, so if adjustment is necessary, several trials will be necessary to complete it.

The miter can be used in either dado groove on the table. When crosscutting a piece to length, be certain to saw on the correct side of the mark. Most saws make a cut about 1/8" wide, so cutting on the wrong side of the mark will introduce an error of about 1/8" in the length.

You may be tempted to cut pieces of lumber to length by setting the fence to the proper distance from the saw blade and allowing the end of the piece to slide along the fence as it is pushed forward with the miter gauge. This is a very dangerous practice. If the piece is inadvertently tilted, as can easily happen because of the friction between the end of the piece and the fence, it will jam on the saw blade. This may result in damaging the saw blade or the piece may be thrown back at you by the saw. You can do this without danger with the arrangement shown diagramatically in Fig. 3.8, where *A* is the fence, *D* is a block of wood clamped to the fence, *C* is the miter gauge head, *B* is the piece of lumber being cut to length and *S* is the saw blade. The piece of lumber is clamped on the miter head in the position indicated in Fig. 3.8, but when it is pushed forward to the saw, the end *d* will not be dragging against anything. It is also

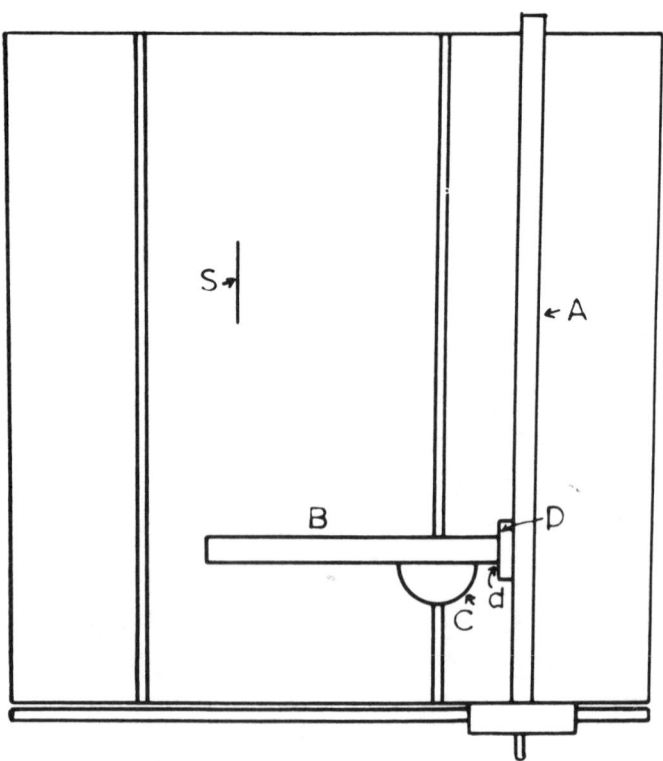

Fig. **3.8   Use of a wood block and the fence for making a series of cutoffs of the same length.**

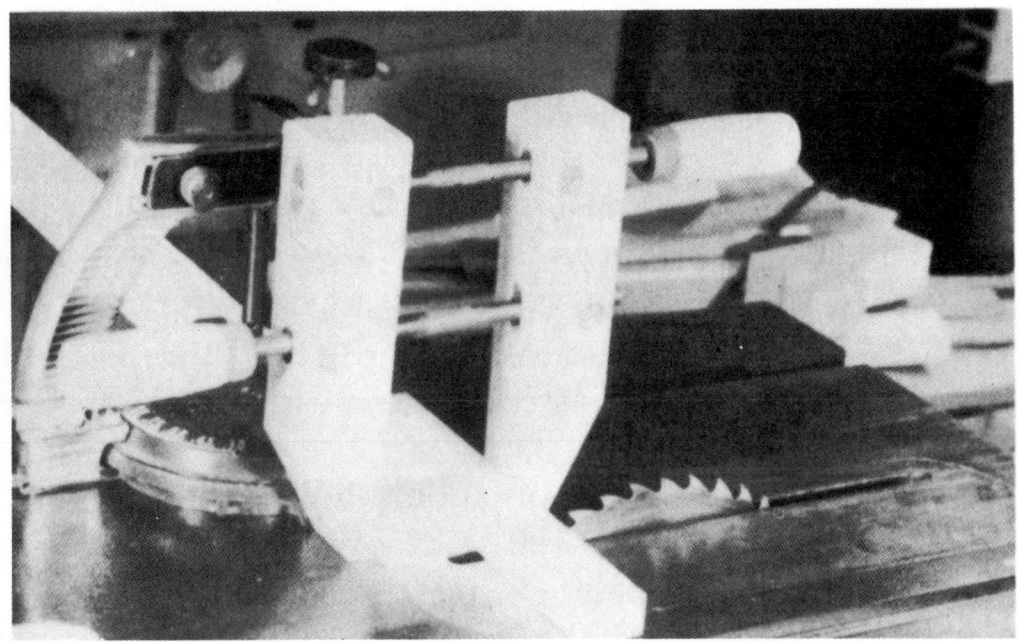

Fig. **3.9   Use of a jig piece and clamp for cutting off short pieces of lumber.**

possible to purchase attachments for the miter gauge which can be set to establish a given length of stock. These are convenient when a number of pieces are to be cut to the same length.

It is sometimes necessary to cut off a portion of a piece too short to be held by the miter gauge. To do this, you can use a jig piece about two inches wide, with parallel edges which can be clamped in the miter gauge so that its end will just miss the saw. The piece to be cut can then be clamped to this jig piece for cutting as indicated in Fig. 3.9. If the piece to be cut is very wide, the miter gauge can be reversed so that the head is forward. It is difficult to handle this arrangement if the miter gauge does not have a clamp for holding the lumber in position. Since it is not possible to hold a piece of lumber against the face of the miter gauge with your fingers to make an accurate angle cut, the piece will need to be rigidly clamped. If your miter gauge does not have a clamp, you can provide a means of clamping pieces for angle cutting. All miter gauges have two holes for attaching a piece of lumber. You can attach a piece of hardwood, such as maple or oak, about 10" long with parallel sides, to the face of the miter gauge and the piece to be angle cut can be clamped to it. While this arrangement makes it possible to rigidly hold the piece to be cut, it is not nearly as convenient as a clamp on the miter gauge.

Before cutting miters, you should adjust the stops for the 45° positions. Follow the same procedure as that for adjusting the stop for the 90° position except use the 45° angle on the combination square for checking the cut after it is made. As a final check, make 45° cuts with the same setting of the miter gauge on two pieces with parallel edges and put them together as indicated in Fig. 3.10 and check with the steel square. When the two mi-

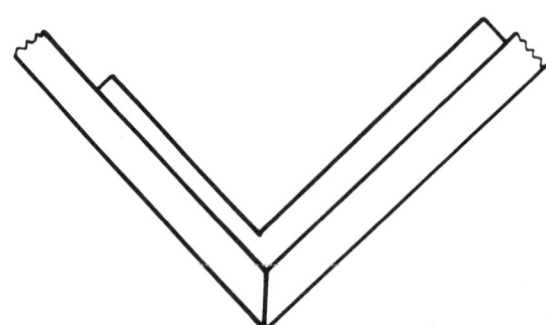

Fig. **3.10 Checking a miter joint for squareness with a square.**

tered surfaces are in contact, the corner should be square.

When you are making a rectangular frame with mitered corners, the miters will not match precisely if the

Fig. **3.11 Setting the angle of a miter gauge with a protractor.**

lengths of opposite sides are not precisely equal, even though the angles of the miters are correct. It is more difficult to cut a piece to a precise length when the cut is at an angle than it is when the ends are cut square. Usually the absolute length of the sides is not as important as the relative lengths of opposite sides. To cut the miters on two opposite sides and insure that they are the same length, the two pieces can be temporarily glued together with two small drops of white glue, one near each end. Let it dry for 15 or 20 minutes with the two pieces clamped in position on the miter gauge, then saw the miters on the two pieces simultaneously; pry them apart immediately and scrape off the glue before it has an opportunity to dry completely.

When cutting angles other than 45°, it is difficult to set the miter accurately using the angle scale on the miter gauge head. This angle setting can be made more accurately by using a protractor to measure the angle between the bar and the face of the miter gauge, as indicated in Fig. 3.11.

3.1.3 The Universal or Tenoning Jig

Figure 3.12 shows a universal jig set up to cut a tenon on the end of a piece of lumber. This is a convenient jig

Fig. **3.12** **Cutting a tenon using the tenon jig.**

for any cutting on the end of a piece of lumber. It consists of a base plate with a bar along the bottom which fits the dado slots in the top of the saw table. This permits accurate movement of the jig from front to back on the saw table. A top plate has a dado slot which matches with a bar on the top of the bottom plate which is perpendicular to the bar on the bottom of this plate. By loosening a large nut above the top plate, it can be moved from side to side for adjustment of the position of the saw cut. The top plate has a vertical portion for holding the work. Angle guides can be mounted in various positions on the vertical portion of the top plate so that the work piece can be oriented accurately vertically and clamps can be attached to hold the work piece rigidly in place while it is being cut.

To cut a tenon, the work piece is clamped to the vertical portion of the top plate with its bottom end in contact with the surface of the saw table and the saw is raised until it will cut a depth equal to the length of the tenon. Its cutting height can be checked by making a trial cut in a piece of scrap stock. The tenon is layed out as indicated in Fig. 2.23 and the top plate is adjusted so that the edge of the saw cut will be on the line marking the edge of the tenon. The cuts A and B in Fig. 2.23 should be made prior to cutting the tenon to avoid splitting out the wood at the ends of the tenon cuts. This jig is useful for holding work for nearly any kind of end cutting on small pieces of lumber. We will later describe molding cutters for use on the table saw. This jig can be used for holding the work piece when a molding is to be cut on the end of the piece.

When a frame is constructed with mitered corners, it is desirable to reinforce the corners after the frame is glued together. One method of reinforcement is to cut a slot through the corner on the edge and glue in a spline. I like to glue the frame together initially without reinforcement and, after the glue has set, add the reinforcement. Figure 3.13 shows the method of holding the frame on the universal jig to cut a slot across the corner with a saw. The saw will cut a slot about 1/8" wide and a piece of scrap stock can be cut to the proper thickness to fit the slot and it can be glued in place and later trimmed down flush with the edges.

This same arrangement can be used for reinforcing a corner of a frame made with a butt joint like that shown in Fig. 3.14. In this instance, the side A will be horizontal and flush with the top of the saw table and the side B will be vertical. The two pieces are held against the vertical portion of the top plate with the clamps while the

Fig. **3.13 Cutting a slot for a spline in a miter joint using the tenon jig.**

cut indicated by the dotted line is being made. The joint is then assembled and glued with the spline in place. The direction of the grain of the wood in the spline should be parallel with the piece *A* for maximum strength. The saw blade used with this jig may be a carbide-tipped saw or better still a special tenon-cutting saw blade which is somewhat thicker than an ordinary

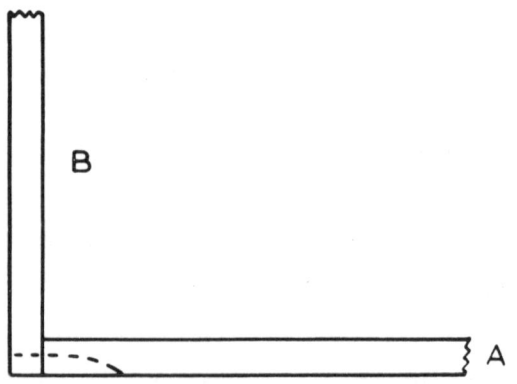

Fig. **3.14 The cut for a spline in a butt right-angle joint, made using the tenon jig.**

saw blade and, therefore, more rigid and less subject to deflection.

3.1.4 Vertical Angle Cutting

The miter gauge is used for cutting lumber at an angle in the plane of the saw table. By tilting the arbor or the table on the saw, it is possible to cut the lumber at an angle in the vertical plane. The range of tilt available is from 0° to 45° in one direction and there is a scale on the saw to indicate the angle of the tilt. Before using this scale to measure an angle, be certain to check the setting of the index by setting the scale to zero and checking the saw with a square. The scale should read zero when the saw blade is perpendicular to the saw table. The saw can be operated in the tilted configuration for either ripping or crosscutting.

3.1.5 Resawing

It is often necessary to produce pieces of lumber thinner than any available in stock. This is accomplished by resawing the lumber. If the piece to be resawed is narrower than the maximum height to which the saw blade can be raised above the table, the process is simply edgewise ripping of the lumber. However, since the lumber rests on its edge on the saw table, it is necessary to depend on the fence to hold the face of the piece vertical. You should check the face of the fence with the small square or the try square to insure that the resawed piece will have the same thickness on the top and bottom.

It is possible to resaw lumber of width up to twice the heighth of the saw blade above the table by sawing through from one edge and turning the piece over and sawing through from the other edge. When resawing such wide pieces of lumber, it is particularly important to be certain that the face of the fence and the saw blade are accurately perpendicular to the saw table if pieces of uniform thickness are to result.

3.1.6   Dado and Molding Cutting

A dado is a rectangular slot cut in a piece of lumber. If no special tools are available, the dado can be cut with a saw blade by making a series of successive cuts with the saw. To do this, the saw must first be set to make a cut of depth equal to the depth of the desired dado. Most saws have a depth-of-cut scale. However, this scale cannot be relied on for determination of the absolute depth of cut because the scale reading, when the saw just contacts the lumber, depends on the diameter of the saw. The scale is useful, however, for determining the change necessary to change from one depth of cut to another. You can determine when the saw is set to the proper depth of cut by checking with a piece of scrap lumber.

The dado can be cut by first setting the fence so that a cut will be made along one edge of the dado. The fence can then be moved an amount equal to slightly less than the width of the saw cut and a second cut is made. This process is repeated until the full width of the dado is cut.

It is easier to cut a dado with the dado cutter shown in Fig. 3.15. This cutter consists of two saws with a series of cutters which can be sandwiched between them. The saws are each 3/32" thick. Four of the supplementary cutters are 3/32" thick and one is 1/16" thick. The dado cutter shown in Fig. 3.15 can cut a dado from 3/16" to 5/8" wide in steps of 1/16".

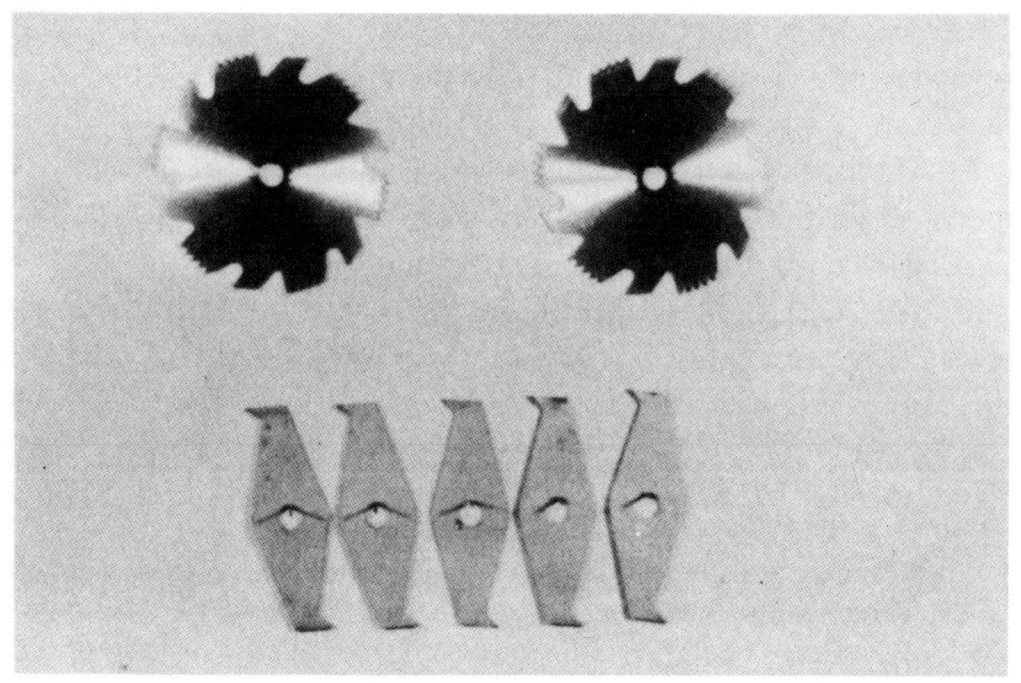

Fig. **3.15   A dado cutter.**

One of the most common moldings is the cove, of which Fig. 3.16 is an example. The cove is cut with the saw. Any blade can be used but it is desirable to use a small diameter blade if one is available.

The width of the cove is indicated at *A* in Fig. 3.16 and the depth is indicated at *B*. The cove is cut by moving the piece of lumber through the saw at an angle.

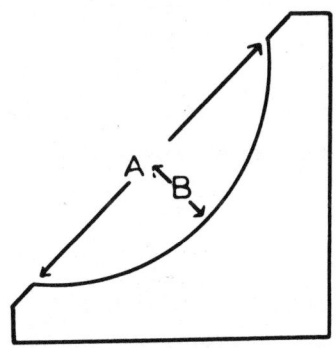

Fig. **3.16   A cove of width *A* and depth *B*.**

The width is determined by the angle and the depth is equal to the height of the saw above the table when the final cut is made.

In order to determine the proper angle at which the piece should be pushed through the saw, a rectangle like that shown in Fig. 3.17 is constructed of scrap lumber so that the width *A* is equal to the width of the finished cove. The saw blade is raised to a height above the saw

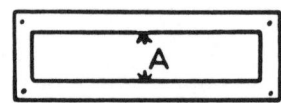

Fig. **3.17 A temporary jig for use in determining the angle of the fence when cutting a cove. *A* is the width of the cove.**

table equal to the depth of the cove and the rectangle is placed over the saw blade and turned to the angle where its two edges just touch the saw blade *S*, as indicated in Fig. 3.18. The sides *a* are parallel to the direction the piece on which the cove is to be cut must be pushed

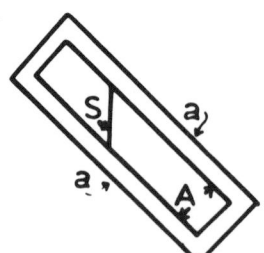

Fig. **3.18 Method of using the jig shown in Fig. 3.17. The saw is set at a height equal to the depth of the cove to be cut.**

through the saw. It is necessary to have a temporary fence on the saw table at this angle and in the proper position to cut the cove in the right place. One way of accomplishing this is shown in Fig. 3.19. The two pieces,

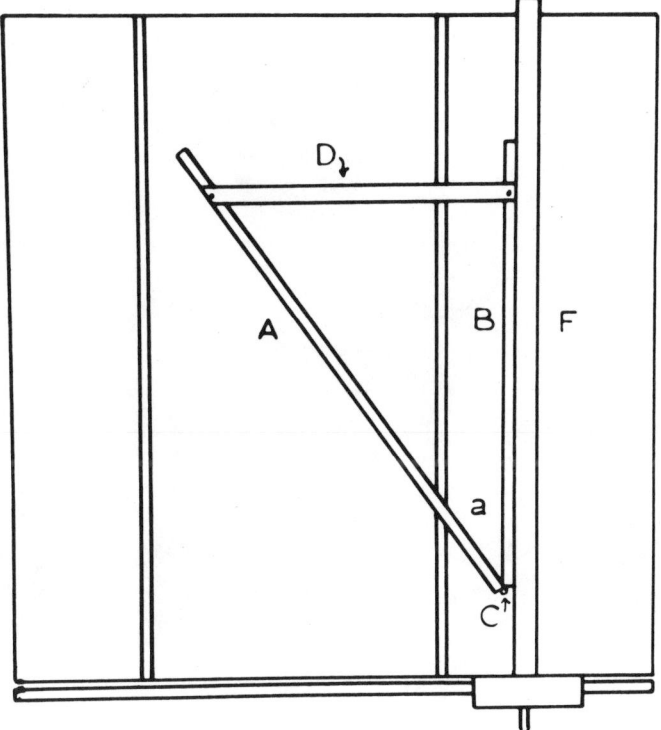

Fig. **3.19 A special fence for use in cutting coves.**

A and *B*, have a hinge at *C* so that the angle *a* can be varied and the strip *D* can be used to hold *A* relative to *B* at the proper angle so that *A* can serve as the temporary fence. In order to have the fence in the proper position to properly cut the cove, *B* can be moved along the saw fence *F* and clamped to it with a C-clamp.

To start cutting the cove, lower the saw blade until it projects about 1/16" above the saw table and, with the piece of lumber against *A*, move *B* along the saw fence until the saw contacts the piece of lumber at about the midpoint of *A* in Fig. 3.16. Push the piece a short distance along the temporary fence and check to determine if the cut is at the center of the cove. You will probably need to make a slight readjustment of the position of the temporary fence. When it is properly adjusted, push the piece through its full length. The saw should be

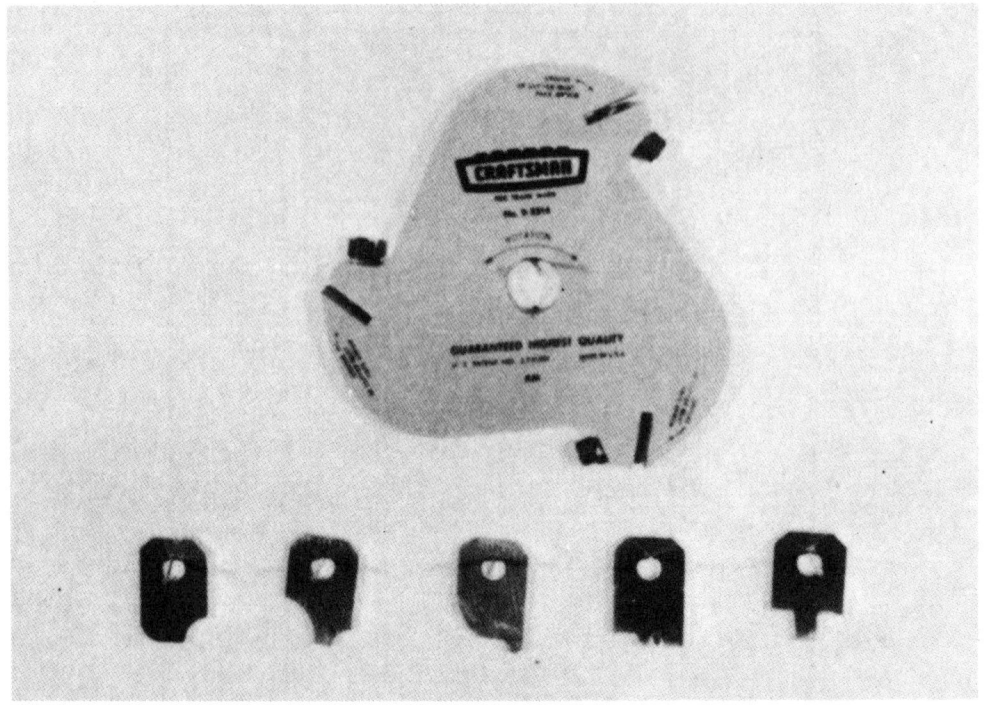

Fig. **3.20 A molding cutter head with some representative molding cutter blades.**

raised a short distance and a second cut made. The amount the saw blade can be raised between successive cuts is determined by the angle *a* and the kind of wood in which the cove is cut. This can be determined by trial. Successive cuts should be made until the cove is cut to the proper depth.

Other types of moldings must be cut with special molding cutters. These cutters must be mounted in a special molding head which is mounted on the saw arbor in place of the saw blade. Figure 3.20 shows a molding head and some representative cutters. Three identical cutters must be mounted in the slots in the molding head and there are about twenty types of cutters available for use in the molding head shown in Fig. 3.20. In some instances, cuts are made with more than one set of cutters to make the complete molding. You should study the molding you wish to cut and work out the combination of cutters you will need to use in order to cut the entire molding. You should not try to cut the entire depth of the molding in a single cut because this will not result in as smooth a cut as can be made with a series of successive cuts.

In order to use the molding head with its cutters, you will need to have a special plate to close the throat in the saw table. This plate must have a shorter and wider slot than the slot in the plate used with a saw blade. If you cannot purchase such a plate, you can purchase a standard plate and cut out the special slot. Do not attempt to cut moldings without a plate to close the throat.

When cutting moldings it is necessary to use the fence to guide the work piece. In some instances, the fence may have to be positioned so that a part of the cutter could strike the metal fence. To prevent this, you should prepare a wood face that can be clamped to the fence. Figure 3.21 shows such a wood face in place on the fence. The cutout on the bottom of the wood face, to

Fig. **3.21 Wood face for the fence for use when cutting moldings.**

prevent interference with the molding cutters, can be made with the dado cutter.

3.1.7 Straightening Lumber

Occasionally a wide board which one wishes to use has some warp or twist which must be removed. This can be accomplished with the saw, using the arrangement shown in Fig. 3.22. The board to be straightened, *B,* is mounted between two strips, *A* and *A',* with nails, *C* and *C'.* The strips, *A* and *A',* should be wide enough so that they will bridge the dado grooves on the saw table.

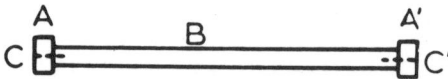

Fig. **3.22**   **Arrangement for use in straightening lumber.**

With the dado cutter on the saw, the unit is placed over the dado cutter and it is raised until it will just make a thin cut on the lowest part of the board *B.* The unit should be guided by the fence and, by moving the fence in steps equal to the width of the dado cutter for successive cuts, the cuts can be made over the entire width of the board. The dado cutter should be raised a small amount and the process repeated until the cutter contacts all of the area of the board. The other side of the board can be straightened in the same manner.

3.1.8 Safety

A table saw must be handled with proper respect. It can become a hazardous instrument if improperly handled. The following is a list of precautions which should always be observed:

1. A piece of lumber to be cut on the table saw should have one straight edge to contact the fence or the miter gauge.
2. Either the fence or the miter gauge should be used to control the piece of lumber being cut. *Never try to cut a piece of lumber freehand.*
3. Never operate the saw with your hands closer to the blade than 4".
4. Never reach over the saw blade when the saw is running.
5. When operating the saw, stand to one side of the direct line in front of the saw blade.
6. Do not allow onlookers to stand behind you when you are operating the saw.
7. Do not wear a long-sleeved shirt, necktie or rings when operating the saw.
8. Be certain to use a three-wire cord for connecting the motor to the electrical outlet, with the ground wire properly connected at each end.
9. When changing blades on the saw, disconnect the electrical plug from the receptacle.
10. Always stop and think before performing any operation with the saw.

3.2 THE RADIAL ARM SAW

The radial arm saw differs from the table saw in that the motor and saw blade are above the working surface. The motor and saw blade are supported on an arm which is pivoted from a supporting column. This arrangement is the reason for the name "radial arm." Figure 3.23 is a photograph of a typical radial arm saw.*

*I am indebted to Dr. J. R. Mentzer for allowing me to make photographs of his saw for Figs. 3.23, 3.24 and 3.25.

Fig. **3.23   A radial arm saw.**

When crosscutting, the saw and power unit are moved in guides on the radial arm and the lumber being cut is fixed on the table. The arm can be rotated to various angles about its pivot at the top of the column to make crosscuts at an angle. Stops are provided at 90° and 45° on each side for cutting miters. These stops perform the same function on the radial arm saw as the corresponding stops on the miter gauge serve on the table saw. The same procedure should be followed in adjusting these stops as that given for adjusting the stops on the miter gauge in Section 3.1.2.

To rip lumber, the saw and power unit are pivoted through 90° in their support on the radial arm and the blade can be moved in the guides on the radial arm to establish the width of the cut. When ripping, the lumber is pushed through the saw in the same manner as when ripping with a table saw. The unit is oriented so that the lumber will feed into the saw with the blade turning so that the teeth on the bottom of the saw turn toward the operator. The depth of cut is adjusted by raising or lowering the support for the radial arm on the column.

All of the operations, such as vertical angle cutting, resawing, dado and molding cutting and straightening lumber are done in a manner similar to that with a table saw. These operations are done with the saw set in the ripping mode and the only difference, really, is that the cuts are made from the top of the lumber rather than from the bottom. There is no equivalent of the tenoning jig for the radial arm saw. Most radial arm saws are so arranged that the power unit can be turned so that the arbor shaft is vertical with the saw arbor at the top. Molding cutters can then be mounted in a receptacle at the bottom end of the shaft.

The radial arm saw is considerably more convenient than the table saw for crosscutting and angle cutting long pieces of lumber because the lumber is fixed on the table while the cut is being made. This type of saw is especially convenient for cutting rafters on construction jobs. A table saw is capable of higher precision work on small pieces of lumber because of the greater rigidity of support of the saw. Unless the radial arm saw is constructed with a very heavy column and radial arm, it is possible for the saw to be deflected slightly when there are sidewise forces such as are encountered in angle cutting.

One disadvantage of the radial arm saw is that the width of a rip cut is limited to about 26" and the maximum width of lumber which can be crosscut is about 15" to 18" on a typical saw.

The same kinds of blades, dado and molding attachments are used in the radial arm saw as are used in the table saw.

3.2.1 Crosscutting and Angle Cutting

The radial arm saw has a fixed fence at the back of the table. The work is placed on the table against the fence with the saw and power unit pushed back, in its guides on the arm, as near to the column as it will go. When the work is in position, held firmly against the fence, the saw is started and the saw and power unit are pulled toward the operator. The saw passes through a slot in the fence and, as it is pulled forward, it cuts its

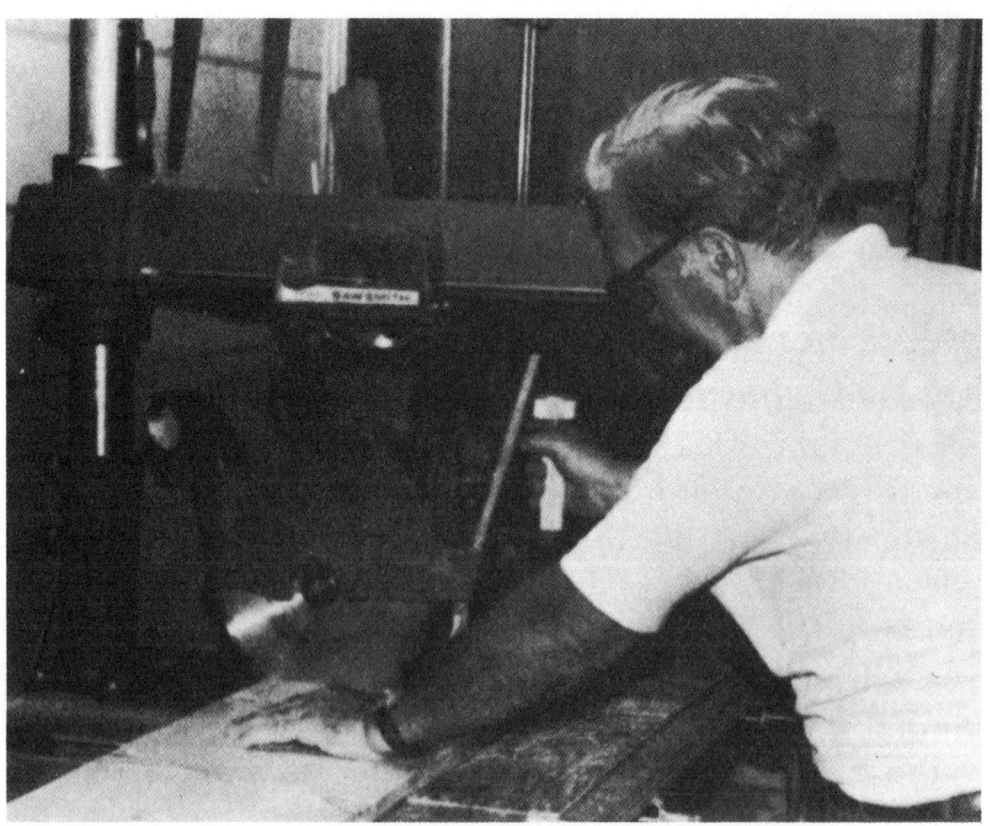

Fig. **3.24 Crosscutting with a radial arm saw.**

way through the lumber. The saw turns in a direction such that the teeth at the bottom of the blade turn away from the operator so that there is a tendency for the saw to feed itself. The handle must be grasped firmly so that the rate of feed will be under control throughout the cut. When crosscutting at an angle, there is a tendency for the lumber to creep in the same manner as with the table saw. This can be prevented by clamping the piece to the fence or table with a C-clamp. Figure 3.24 shows a piece of lumber being crosscut with a radial arm saw.

3.2.2 Ripping

To rip a piece of lumber, rotate the saw and power

Fig. **3.25 Ripping with a radial arm saw.**

unit through 90° in its support on the arm. Since the fence is fixed, the unit is moved in its guides on the arm to the proper position to rip the desired width. The width of the rip can be set in the same manner as that indicated in Fig. 3.4. The lumber is then pushed through the saw, keeping the edge firmly against the fence. Figure 3.25 shows a piece of lumber being ripped on a radial arm saw. The same procedure is followed in cutting dadoes and moldings and in resawing.

1. A piece of lumber to be cut on a radial arm saw should have one straight edge to contact the fence.

2. The fence should always be used to control the piece of lumber being cut. *Never try to cut a piece of lumber freehand.*

3. Never operate the saw with your hands closer to the blade than 6".

4. Always return the saw to the rear of the table after completing a crosscut.

5. Shut off the motor and wait for the blade to stop before making adjustments.

6. When ripping, always feed the stock into the blade so that the bottom teeth are turning toward you.

7. Do not wear any loose clothing such as longsleeved shirts or neckties when operating the saw.

8. If you have long hair, put it up in a net or cap.

9. Always be certain that the saw guards are in place and working properly before using the saw.

10. When crosscutting, be certain that the antikickback device is in place and functioning properly.

3.3 THE HAND POWER SAW

The hand power saw is like a miniature table saw upside down. A metal base plate serves the function of the table and the amount of projection of the saw blade through this plate is adjustable. The blade is driven directly by the motor and a handle on the motor frame is provided for operating the saw.

There is an index at the front of the base plate that can be used to follow the line along which the cut is to be made. A spring-loaded guard surrounds the portion of the saw blade that projects below the base plate. When a cut is to be made, the front of the base plate is placed on the lumber to be cut and the saw is pushed forward, keeping the index on the line to be cut. When the guard contacts the edge of the piece of lumber, it is pushed back against the spring, allowing the saw to cut the lumber. When the saw passes beyond the far edge of the piece of lumber, the guard snaps back, covering the blade again.

These saws are used primarily by carpenters. A skilled carpenter can follow a line very accurately with one of these saws but an amateur who would use such a saw only occasionally will have difficulty cutting a straight line.

The primary use for such a saw by an amateur is in cutting plywood, particularly in cabinetmaking. Plywood sheets are so large that they are very difficult to handle on a table saw or a radial arm saw. They can be cut easily and quickly with a hand power saw by placing the plywood on sawhorses and setting the depth of cut of the saw approximately 1/16" greater than the thickness of the plywood. The saw cuts can then be made right across the tops of the sawhorses.

Cuts on plywood can be made with great accuracy by using a temporary fence consisting of a straight piece of lumber that can be clamped to the plywood at a distance from the line of the cut equal to the distance from the saw blade to the edge of the base plate. It is, of course, necessary to determine, in each instance, if the distance should be measured to the near or far side of the saw blade. Figures 2.5 and 2.6 in *Amateur Cabinetmaking* by Vernon M. Albers show the details of the use of a temporary fence with a hand power saw.

The hand power saw uses blades like those used in the table saw and the radial arm saw, except that they are smaller in diameter. The blades used are usually 6" or 7" in diameter. Since carpenters often work with lumber which can damage the saw blades, it is possible to purchase "throwaway" blades which are quite inexpensive but cannot be resharpened. When one of these blades is damaged or becomes dull it is simply discarded and replaced with a new one.

There are several grades of hand power saws available. The cheaper ones are lighter in weight than the more elaborate models but are completely adequate for use by an amateur. An amateur who only occasionally needs to cut plywood does not need one of these saws since it is possible to do the cutting with a hand saw. However, if he intends to construct a number of cabinets of plywood or chipboard, such a saw can save a great amount of physical labor.

3.4 THE BAND SAW

The band saw, as its name implies, has a saw blade in the form of a band that runs around two wheels. One of the wheels is driven by a motor and the other is an idler. The wheels have a tire of a rubberlike material on which the band rides. The idler wheel is usually at the top and its tilt relative to the bottom wheel is adjustable to insure that the band will run on the approximate centers of the tires. This adjustment is usually made with a nut on the axis of the wheel. Turning the nut tilts the wheel. There is a lever which raises the upper wheel to tighten the band and when bands are changed it is necessary to release this lever to loosen them.

You can check to determine if the band will run on the center of the wheel tires by turning the wheels by hand with the front guard removed from the saw. It is necessary to remove this guard to change blades. Do not operate the saw with the motor when the guard is not in place. Figure 3.26 is a photograph of a band saw and Fig. 3.27 shows the saw with the guard removed.

Blades for wood cutting are available in widths from 1/8" to 3/4". Most of the cutting with the band saw is done with a 1/4" or a 3/8" blade. Metal-cutting blades are also available but it is necessary to have a speed reducer when cutting metals. There are blades for cutting soft metals, such as aluminum, and also blades for cutting ferrous metals. The speed of the blade for metal cutting is usually about one-tenth that for wood cutting.

Some band saws have the speed reducer built in but for those which do not have a built-in speed reducer, it is necessary to purchase a special unit which can be mounted on the motor shaft if the saw is to be used for metal cutting. These speed reducers have two pulleys, one which rotates at the speed of the motor shaft and the other which rotates at the slower speed. It is simply necessary to shift the belt from one pulley to the other to change the speed of the band.

There are two guide units for the band, one above the table and one below the table. Figure 3.28 shows the arrangement of these guides. The wheel *W* is at the back of the band, the side opposite the teeth, and the position of this wheel is adjustable to accommodate different

Fig. **3.26 A band saw.**

widths of bands. With the band running freely, the back of the band should just touch the wheel and, as a cut is being made, the back of the band rides against this wheel. The two side guides G should be adjusted so that when the blade is running freely it does not ride against

Fig. **3.27   A band saw with the front guard removed.**

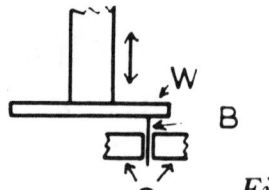

Fig. **3.28** **The guides for a band saw blade.**

either of them. There should be about 1/32" clearance between each of these guides and the band when the saw is running freely. The wheel *W* should be adjusted so that the teeth on the band will not strike the guides *G* when the blade is pushed to one side or the other when cutting curves.

The guide unit below the table is fixed in position but the one above the table can be raised or lowered to accommodate different thicknesses of lumber. The upper guide unit should be set so that there is approximately 1/4" clearance between the top of the material being cut and the bottom of the guide unit. This setting is particularly important when thick materials are being cut. There is a guard around the top guide unit which must be removed when the unit is adjusted and when bands are being changed. Do not do any cutting with the saw when this guard is not ín place.

The bands for band saws normally used by amateurs are relatively inexpensive and should be replaced with new ones when they break or become dull with use.

3.4.1 *Crosscutting with the Band Saw*

Crosscutting can be done either freehand or with the work held by a miter gauge. You can usually use the miter gauge from your table saw on the band saw as the size of the slot for the bar of the miter gauge is standard on most saws. When cutting free hand, the line of the cut should be marked and the work guided to follow the mark with the cut. With the miter gauge, 90° cuts and angle cuts are made in the same manner as with a table saw.

With wide pieces, a 90° cut can be made using a fence. If you do not have a fence for your band saw, a temporary fence can be improvised by clamping a straight piece of lumber to the table at the proper distance from the band blade.

3.4.2   Ripping and Resawing

Ripping can be done freehand by following a line on the piece to be cut or the piece can be ripped using a fence in the same manner as a 90° crosscut is made.

It is possible to resaw wider boards on a band saw than on a circle saw. However, if you have a circle saw, you should cut a groove at least 1" deep from each edge with the circle saw and then do the remainder of the cut on the band saw. The grooves cut with the circle saw will help to guide the band saw blade. It is, of course, necessary to use a fence when resawing and you should use the widest blade available.

3.4.3   Other Operations with the Band Saw

The important advantage of the band saw over circular saws is that it is possible to cut curves with it. The minimum diameter of a curve that can be cut depends on the width of the band used. Table 3.1 gives the minimum diameter of curves that can be cut with the various widths of bands. The cuts should be planned so that the curves can be cut with minimum strain on the blade. Figure 3.29 shows how supplementary cuts can be made to minimize the strain on the blade when the

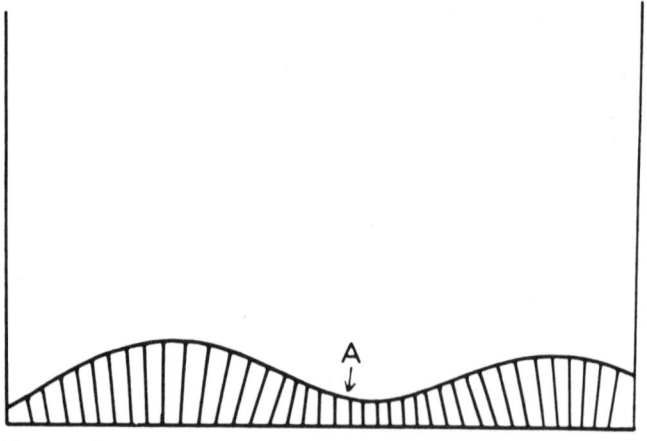

Fig. **3.29   Use of supplementary cuts to minimize the strain on a band saw blade when cutting curves.**

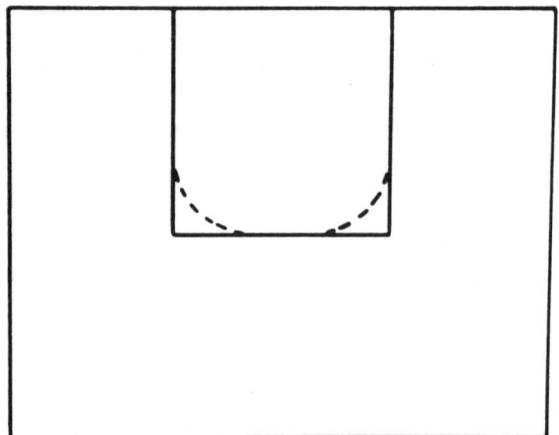

Fig. **3.30   Method of making a rectangular cutout with a band saw. The first cut is made on the dotted line and then the corners are cut out.**

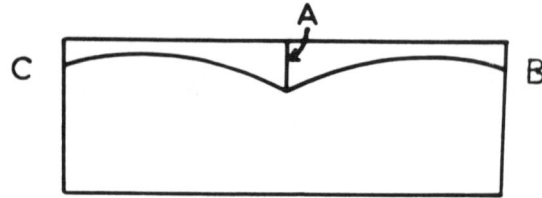

Fig. **3.31   Method of cutting a curve with an inflection. *A* is a supplementary cut.**

curve *A* is cut. The supplementary cuts are made from the outside edge of the piece just to the line of the curve to be cut.

When a right angle cutout is to be made, the cut can first be made along the dotted line in Fig. 3.30 and the waste wood in the corners can be cut out by cutting up to the corners from each direction.

When a curve with an inflection like that shown in Fig. 3.31 is to be cut, a supplementary cut *A* is first made and the piece is then cut from *B* to the cut *A* and then from *C* to the cut *A*.

In cutting curves you should always plan your operation so that you will impose the minimum strain on the band. On the basis of Table 3.1 you can select the appropriate band width for the particular job to be done. The 1/8" band is quite fragile and should not be used unless very small radius cuts are to be made. When it is necessary to back the blade out of a cut, the saw should first be stopped.

Table 3.1.
Minimum diameter of cuts for different band widths

Band width	Diameter
1/8"	1/2"
1/4"	2"
3/8"	3"
1/2"	5"

3.4.4 Safety

1. Never operate the band saw without the front guard in place.

2. Do not do any sawing without the guard around the top guides in place.

3. When resawing wide pieces of lumber, always saw slots from the two edges with a circle saw to serve as guides for the band.

4. When sawing freehand, avoid making abrupt changes in direction of the cut, which could jam the blade.

5. When it is necessary to back the blade out of a cut, the saw should be stopped.

6. Avoid wearing loose clothing when using a band saw and if you have long hair, put it up in a net or cap.

3.5   THE JIG SAW AND SABER SAW

The band saw cannot be used for cutouts that do not extend to the outside edges of the piece.

The jig saw is, essentially, a motor driven scroll saw. It uses blades similar to scroll saw blades and the blades are driven vertically through a horizontal table by means of a cam on a shaft driven by a motor. If an internal cutout is to be made, a hole can be bored in the piece to be cut and the blade can be assembled in the saw, through the hole, and the cut is then made in a manner similar to making a cut with a band saw. The most common use of the jig saw is in making small radius cuts in thin materials. Jigsaw puzzles get their name from the fact that they were originally cut with jig saws.

A jig saw is not often found in an amateur's shop. It is not as effective as a band saw where a band saw can be used and, for internal cutouts, a saber saw is usually more convenient because it is completely portable and does not require any bench space for mounting.

The saber saw uses a short, stiff blade driven by a cam from a small motor. It is used for essentially the same operations as the jig saw. A variety of types of blades are available for wood cutting and metal cutting. It is possible to purchase a complete saber saw or you can purchase a saber saw attachment for use with some hand electric drills so that the motor in the hand electric drill is used to drive the cam which operates the saber saw blade. Figure 3.32 is a photograph of a 1/4" hand electric drill with a saber saw attachment.

Fig. **3.32   A saber saw attachment on a hand drill.**

3.6   SHARPENING CIRCLE SAW BLADES

Circle saw blades are sharpened in much the same manner as hand saw blades. The principle difference is that circle saw blades have larger teeth and the teeth are arranged around a circle instead of in a straight line.

The first step is to joint the saw blade. With a table saw, the blade is first lowered below the surface of the table and, with the saw running, place a coarse Carborundum stone on the table over the saw blade and gradually raise the blade until you can see sparks as the teeth contact the stone. Inspect the blade frequently and continue the operation until all of the teeth have been jointed. The blade is now ready for filing. The process of jointing a blade on a radial arm saw is similar to that with a table saw except that the blade is lowered to the stone from above. When jointing a saw blade on a radial arm saw you should wear safety glasses.

To file the teeth, the blade should be clamped between circular wood discs that are smaller in diameter than the saw blade and the unit, consisting of the saw blade and the wood discs, can be clamped in a vise for convenience in filing.

Most combination blades which are used by amateurs have teeth similar to the teeth on rip saws (see Fig. 1.6). The blades *B* and *C* in Fig. 3.2 have periodically spaced large spaces between teeth to help in carrying out the saw dust. Usually the small teeth are the cutting teeth and the large teeth are slightly shorter and serve merely to wipe out the saw dust.

On most saws, the teeth should be filed like those on a hand rip saw, that is, with the file stroked nearly perpendicular to the surface of the saw blade. Since the teeth on a circular saw are larger than the teeth on a hand saw, the file used should be larger. Some saws are made of steel so hard that the files are rapidly dulled.

The blades with carbide-tipped teeth, like that shown at *A* in Fig. 3.2, cannot be sharpened by the user. When these blades become dull, they shoul.d be sent back to the factory for reconditioning. The blade *D* in Fig. 3.2

has teeth comparable in size to the teeth on a hand saw and they should be sharpened in the same manner as the teeth on a hand saw. Before starting to file the teeth on any circular saw blade, carefully examine them to determine the angle at which the file should be stroked.

The blade *B* in Fig. 3.2 is a tapered blade and the teeth on such a blade are not set because the teeth are at the thickest part of the blade. Blades which are not carbide-tipped or tapered have teeth which should be set. The principle involved in setting the teeth on a circle saw is the same as that involved in setting the teeth on a hand saw. However, the larger teeth and the heavier blade of the circular saw make it necessary to use a different kind of saw set. Figure 3.33 shows a saw set being used to set the teeth on a circle saw blade. There are two parts to the saw set. One part contains a

Fig. **3.33 A saw set for use on a circle saw.**

small arbor on which the saw is mounted and the other part contains the anvil against which the tooth is set and a hard steel, spring-loaded pin that can be driven against the tooth to set it. These two portions of the saw set are designed to be clamped in one of the grooves on the saw table. The part that holds the saw blade is first clamped in position and the saw blade is mounted on it and the part that contains the anvil is clamped in position so that the teeth fall in the correct position to be set. The tooth is set by driving the pin on the saw set with a machinist's hammer. The blade at *D* in Fig. 3.2 can be set with the saw set used for setting the teeth on hand saws.

3.7 SHARPENING MOLDING CUTTERS

Figure 3.34 is a cross section of a typical molding cutter. These cutters have a smooth ground side *A* and a bevel *B* on the opposite side. To sharpen these cutters, the smooth side *A* should be placed on a fine Carborundum stone and the blade moved back and forth until the edge *C* is sharp. Do not attempt to grind the bevel *B*; this will change the length of the cutter. It is important for the three cutters in a set to be identical and it is not

Fig. **3.34   A cross section of a typical molding-cutter blade.**

possible to freehand grind the cutters without changing the relative shapes of the cutting edges. If a cutter becomes damaged, the set should be replaced.

4

THE JOINTER

4.1 INTRODUCTION

The jointer is an important tool in the woodworking shop. It consists of a table similar to the bottom of a plane, except that it is considerably larger, with a rotating cutter which performs the same function as the cutting iron in a plane. The jointer differs from a plane in that the table is in two parts, one on each side of the cutters, that can be separately adjusted in level while the bottom of a hand plane is a single piece and the cutter iron is adjustable.

Figure 4.1 shows schematically the arrangement of the parts of a jointer. A is the infeed table over which the work is advanced to the cutter in the direction of the arrow and B is the outfeed table. Both tables are adjustable in the ways C and D. The table B is adjusted so that its level is precisely equal to the level of the highest

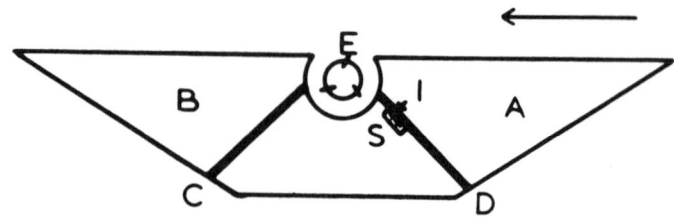

Fig. **4.1** **Arrangement of the parts of a jointer.**

point reached by the cutting edges of the blades in the arbor E. This can be assured by placing a square on the table B so that its end overlaps E and adjusting the level of B so that as E is slowly rotated the blades just touch the square. This adjustment should be left fixed in future operations of the jointer. However, its adjustment should be checked occasionally to be certain that it remains correct.

The table A is adjustable to determine the depth of cut when a piece of lumber is moved along the tables in the direction of the arrow. The depth of cut can be read on the scale S with the index I. The index should be adjusted to read zero on the scale when A is adjusted to be in the same plane as B. This is done by the same method as the adjustment of B. When the index is properly set, the scale reading will indicate the depth of cut, which can usually be varied from zero to 1/2".

In order to be certain that the jointer will plane a straight edge on a piece of lumber, it is necessary for the top surfaces of A and B to be parallel. If the surfaces are out of parallel, as indicated in Fig. 4.2, where the dotted line A is a straight line, the jointer will plane an edge with a concave curvature; if they are out of parallel, as indicated in Fig. 4.3, the jointer will plane an edge with a convex curvature. The misalignment indicated in Figs. 4.2 and 4.3 is, of course, greatly exaggerated. If the

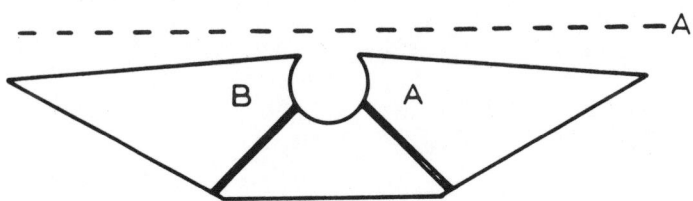

Fig. **4.2 Misalignment of the tables of a jointer that will produce a concave cut.**

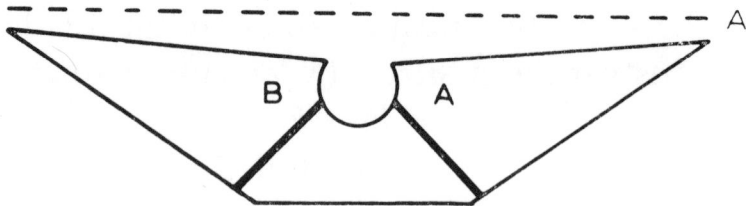

Fig. **4.3 Misalignment of the tables of a jointer that will produce a convex cut.**

jointer planes edges which are not straight, the alignment of the two tables should be checked with the large steel square. Some adjustment is possible with the metal strips which clamp the tables in the ways, but it is sometimes necessary to completely disassemble the jointer and clean the ways, as an accumulation of dust may introduce some error of tracking in them.

A 1/2-horsepower motor is adequate to operate 4" and 6" jointers. Some amateurs prefer to purchase a special stand which can accommodate both a table saw and a jointer using a single motor to operate both units. This arrangement is not satisfactory unless the saw is a tilting table type, because a tilting arbor saw requires that the motor be tilted as the arbor is tilted.

Power planes are available which are like a miniature jointer upside down. These planes are not capable of doing as precise work as a jointer because of their small size. Their primary application is for work where a portable unit is required.

4.2 PLANING AN EDGE

The jointer has a fence, as indicated in Fig. 4.4, where *T* is the table and *F* is the fence. The fence can be tilted

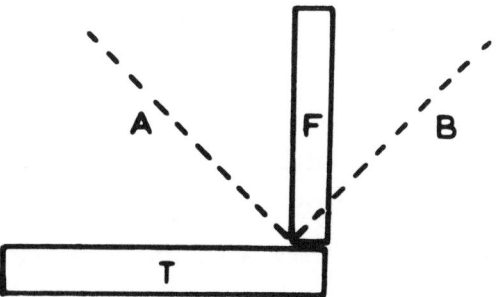

Fig. **4.4 The relation of the fence to the tables of a jointer.**

from the + 45° position to the − 45° position, indicated by the dotted lines *A* and *B*. There are stops at the zero position, where the fence is perpendicular to the table, and the two 45° positions. These stops can be adjusted by the method described in Section 3.1.2. There is a spring-loaded guard over the cutters which is pushed aside as the work is pushed through the cutters.

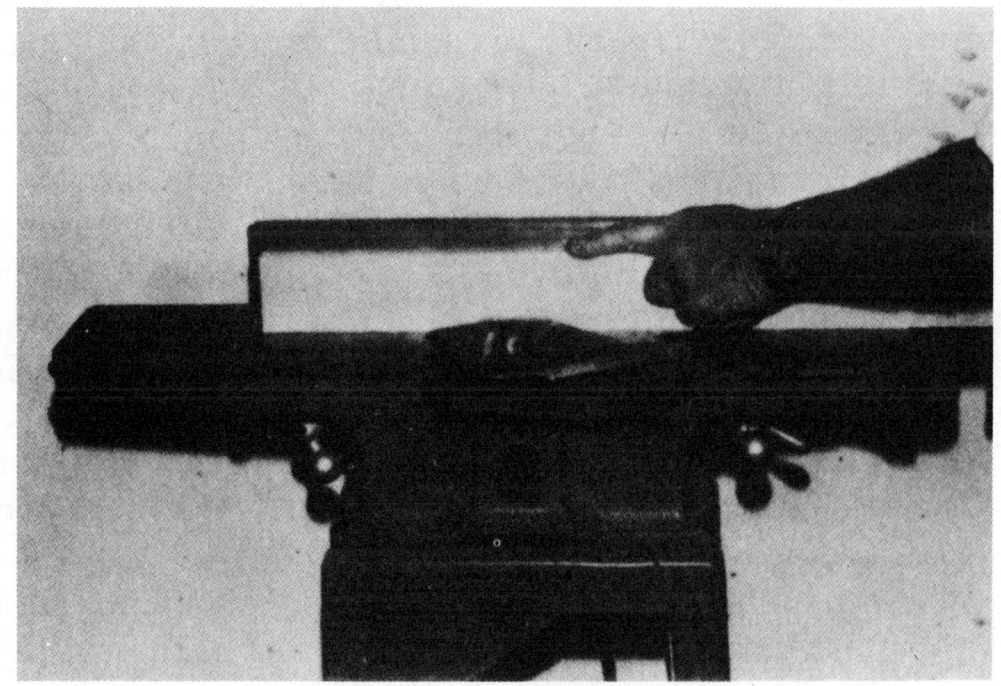

Fig. **4.5 A piece of stock being edge-jointed.**

When planing an edge, examine the stock to determine the direction of the grain. It should be fed through the jointer knives so that they will not cut against the grain to avoid chipping. Figure 4.5 shows a piece of stock being edge-jointed. Place the stock on the infeed table, hold it firmly against the fence, and slowly push it against the cutters. Do not make a cut of more than 1/8". If the edge of the piece is being planed to form a part of an edge-glued joint, the final cut should be thin, not over 1/32". Note in Fig. 4.5 how the guard is pushed aside by the work piece as it passes over the cutting knives. As the piece passes out over the outfeed table, the guard snaps back into its position over the knives.

4.3   PLANING A SURFACE

You should turn your piece of lumber so that the knives will not cut against the grain, to avoid chipping. The left hand should be placed on top of the stock, back of the knives at the beginning of the cut and stepped over to a point ahead of the knives as the cut progresses. The right hand is used to push the stock through and, as the knives are approached at the end of the cut, a push stick similar to that shown in Fig. 3.6 may be used for the remainder of the cut. A push block, like that shown in Fig. 4.6, is more effective than a push stick. The block should be made from a piece of 3/4" lumber about 2½" wide and a piece of 5/8" or 3/4" dowel pin stock may be used for a handle.

When planing thin pieces, there is danger of the end slipping under the fence, where there is clearance between the fence and the outfeed table. To avoid this, a

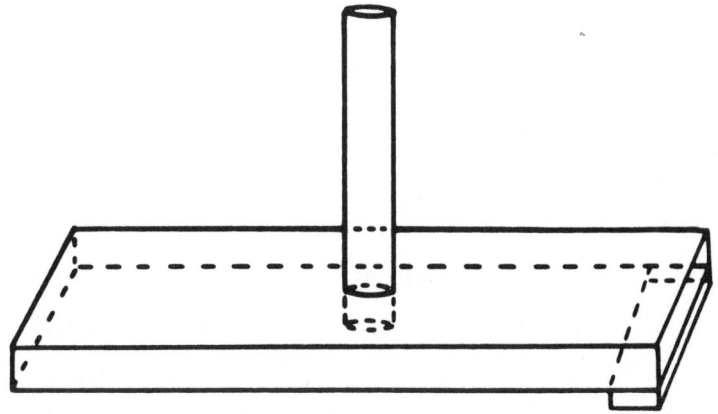

Fig. **4.6** **A push block for use in surface planing.**

Fig. **4.7** **Planing a bevel with a jointer.**

board should be clamped to the fence so that its bottom edge is in contact with both tables.

4.4 PLANING A BEVEL

The fence can be tilted in either direction to plane a bevel. The fence is set at the proper angle and the piece is planed as indicated in Fig. 4.7. It is usually easier to hold the piece to be planed firmly against the fence if the fence is tilted toward the tables because there is a tendency for the bottom edge of the piece to slip away from the fence when it is tilted away from the tables.

Fig. **4.8   The top of a typical small jointer.**

4.5 PLANING A RABBET

Figure 4.8 shows the top of a typical jointer with the guard removed. The jointer can be used to cut a rabbet by moving the fence to the left so that the distance from the end of the cutters to the fence is equal to the width of the rabbet and the maximum depth of the rabbet will be equal to the maximum depth of cut to which the infeed table can be set. The full depth of the rabbet should not be cut in a single pass. Successive cuts of not more than 1/8", depending on the hardness of the wood, should be made, with the final cut being about 1/32". Figure 4.9 shows a rabbet being cut. Note the

Fig. **4.9   Cutting a rabbet with a jointer.**

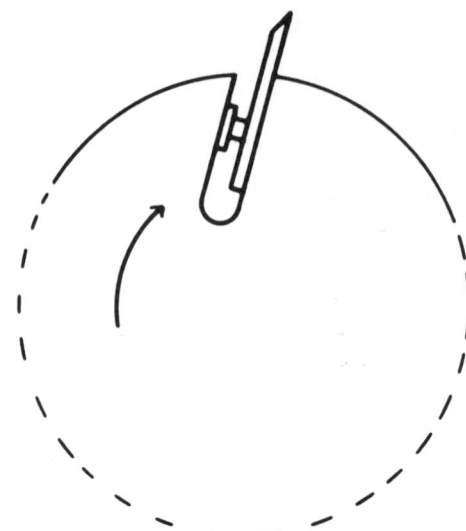

Fig. **4.10   A portion of the arbor of a jointer showing one of the three blades.**

portion of the infeed table which extends beyond the cutters on the outside of the outfeed table. This furnishes support for the piece of lumber outside the rabbet to prevent gouging the cut at the end of the rabbet. The guard must be removed when cutting rabbets.

4.6   SHARPENING JOINTER BLADES

It is necessary for the jointer blades to be sharp to do acceptable work. Figure 4.10 shows schematically a portion of an arbor with one of the three cutters. The arrow indicates the direction of rotation of the arbor. The bevel on the cutters is usually hollow-ground and the cutters can be honed with a small square arkansas stone by placing it on the bevel, with the arbor on the jointer clamped so that it cannot rotate. The stone is stroked back and forth along the length of the cutter.

An alternate method of honing the cutters is to wrap

a piece of paper around a Carborundum stone so that a portion of one end of the stone projects beyond the paper. The portion of the stone wrapped with the paper is placed on the outfeed table with the uncovered portion projecting over the arbor. The outfeed table is lowered and the arbor is rotated so that the cutting edge of the blade touches the stone with the back of the bevel just clearing the stone. The stone is stroked back and forth along the length of the blade. The fence must, of course, be removed while the blades are being honed.

Most amateurs are not equipped to grind a new surface on a jointer cutter blade because a special jig is required to insure that the blade-cutting edge is straight. If the blades are nicked or become so dull that they can no longer be honed to a satisfactory edge, they should be replaced. In most instances a new set of blades should be purchased, but some manufacturers have a service whereby a second arbor can be purchased, complete with blades properly mounted. When a set of blades needs to be resharpened, the arbor can be replaced with the spare and the one with the dull blades can be sent back to the factory, where the blades will be sharpened and remounted, ready for use.

If it becomes necessary for you to replace a set of blades, the proper alignment is very important. The first step is to align the tables, with the infeed table set with the index at zero on the scale. If the jointer does not have any special arrangement for aligning the blades, it can be done by clamping the rotor in the orientation where the cutting edge is at its highest point and adjusting the height of both ends of the blade so that the edge will just touch a straight edge which is so placed that it spans the gap between the two tables. The arbor can then be firmly clamped. The blade is raised until it just touches the bottom of the straight edge, first on one end and then on the other. It usually requires several trials to achieve perfect alignment, and the final test after the blade is firmly clamped is to turn the arbor shaft back and forth by hand with the straight edge first at one end and then at the other to determine if the blade just touches the straight edge as it passes its highest point. All three blades must be at the same height so that they will contribute equally in cutting and both ends of each blade must be at the same height to insure that the planed edge will be square when the fence is set at an angle of 90° with the tables.

4.7 SAFETY

1. Make adjustments of the position of the fence and depth of cut before turning on the machine.
2. The maximum cut for jointing an edge is 1/8"; for a flat surface, 1/16".
3. The stock to be planed should be at least 12" long and you should avoid surface planing stock less than 1/4" thick.
4. Use a push stick when planing a flat surface. Do not apply pressure directly over the knives with your hand.
5. Do not plane end grain unless the board is at least 12" wide.
6. When honing or adjusting the knives, disconnect the jointer from the power circuit.

5
POWER BORING TOOLS

5.1 THE DRILL PRESS

The most important power boring tool is the drill press. It consists of a base that may be mounted on a bench or on the floor, a vertical column, the .drill table and the mechanism which includes the motor, speed reducer and spindle. If the drill press is to be mounted on a bench, a short column is used, but if it is to be mounted on the floor, a long column must be used.

Since a considerable range of speeds must be available, a means of varying the spindle speed must be provided. The simplest arrangement for speed control is by means of pulley units that consist of a set of pulleys of different diameters cast as a unit. One set is mounted on the motor shaft and the other is mounted on the shaft that drives the drill press spindle. One of these units is mounted with the small pulley at the top and the other is mounted with the small pulley at the bottom. By shifting the belt from one pulley to another the spindle speed can be changed from less than the speed of the motor to greater than the speed of the motor. A much more convenient system consists of an arrangement which includes two idler pulleys whose relative diameters can be varied with a lever to provide spindle speeds from about 150 rpm to about 3,000 rpm.

The spindle is free to slide up and down on the shaft that drives it and the chuck that holds the drills is mounted on the bottom end of the spindle, usually by means of a standard taper with a threaded collar to draw the taper into its socket. The chucks on most drill presses used by amateurs will handle drill shanks from about 1/32" to 1/2".

The drill press table is supported by the column and can be moved up and down on the column and be clamped to it to accommodate the piece of material to be bored at the proper height relative to the end of the drill. Some drill press tables have provisions for rotating the table on its support on the column through an angle of up to 45° on either side. This arrangement makes it possible to bore holes at an angle. Figure 5.1 is a photograph of a bench-mounted drill press.

5.2 DRILLS FOR USE IN THE DRILL PRESS

The drills most commonly used in the drill press are standard twist drills. These drills get their name from the shape of the grooves or flutes which carry out the waste from the hole being bored. Figure 5.2 is a photograph of a twist drill and Fig. 5.3 is a sketch showing the different parts: *a* is the body, *b* is the land, *c* is the spiral groove and *d* is the angle of the cutting edge.

The diameter at the lands is the diameter of the hole bored by the drill and the body of the drill is relieved to reduce the friction in the hole. Figure 5.4 is a drawing of the end of a twist drill. It is important for the two ground surfaces to cut to the same depth and when this is true, the line *a* in Fig. 5.4 will make an angle of about 45° with a line parallel to the two cutting edges.

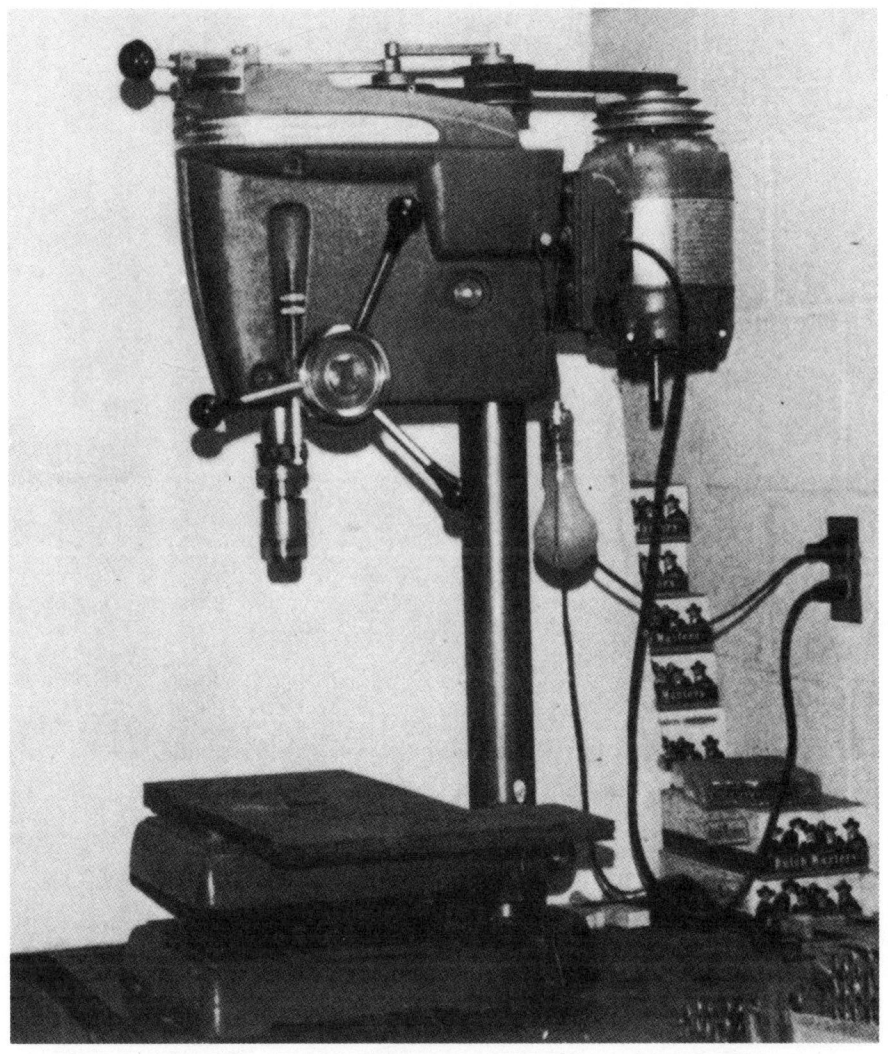

Fig. **5.1   A bench-mounted drill press.**

The normal twist drill sets contain drills from 1/16" to 1/2" in steps of 1/64" and a wire gauge set from No. 1 (0.225") to No. 60 (0.036"). It is also possible to obtain twist drills larger than 1/2" with shanks machined down to 1/2" so that they can be used in a 1/2" chuck. The drill shown in Fig. 5.2 is a 1" drill with a 1/2" shank. When drills smaller than No. 60 are used, it is necessary to mount a supplementary small chuck in the drill press chuck because the standard chuck will usually not hold drills smaller than No. 60.

Fig. **5.2   A large twist drill.**

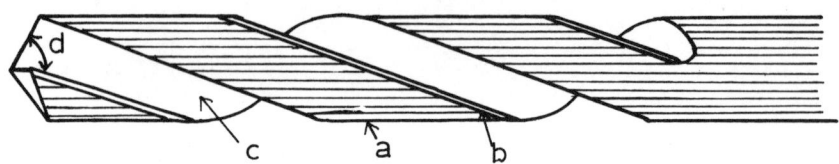

Fig. **5.3   A twist drill.** *a* **is the body,** *b* **is the land,** *c* **is the spiral groove and** *d* **is the angle of the cutting edge.**

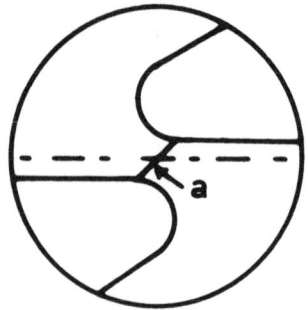

Fig. **5.4   The cutting end of a twist drill.**

It is obvious from Fig. 5.4 why a twist drill may not bore a hole in wood around the desired center. The common method of circumventing this difficulty is to start with a good center punch mark and bore a small hole followed by boring with larger drills until the desired hole size is reached.

The drill shown in Fig. 2.19 is an example of a special drill for wood boring that will bore a hole about the desired center more accurately than a standard twist drill. It is not possible to obtain these drills in all sizes but if one wishes to drill a hole of different size than those available in this type of drill, you can bore the next smaller size with one of these drills and then use a standard twist drill to enlarge the hole to the desired size.

Twist drills are designed for boring holes in metal. When boring in metal, the metal is quite effective at conducting away the heat. Wood is a poor conductor of heat, so care must be exercised in boring in wood with a twist drill. It is easily possible to generate enough heat, when boring wood, to overheat the drill, so it should frequently be raised out of the hole to clear the flutes and allow the drill to cool. Large drills should be run at slower speeds than small drills and, in general, the speeds of drills in boring wood should be less than the speeds for the same drills when boring metal. Twist drills are available in carbon steel and high-speed steel. Either kind can be used for wood boring but the high-speed steel drills do not need to be resharpened as often.

There are other special wood-boring bits for use in a drill press somewhat similar to the one shown in Fig. 2.19, except that instead of the brad point there is a threaded feed screw similar to that on a standard auger bit. These bits are rather expensive so they are not usually found in an amateur's shop. It is also possible to purchase special drills for boring screw holes. These drills have the portion for boring the correct size hole for the screw thread followed by a portion to drill the clearance size for the shank of the screw which, in turn, is followed by a portion which cuts the countersink for flat head screws. These drills are convenient when holes for a large number of screws are to be bored.

There is another type of drill, called a centerpoint drill, which is good for accurately starting a hole in either wood or metal. Figure 5.5 is a drawing showing how such a drill is constructed. With two or three sizes of these drills it is possible to start holes accurately, and if a centerpoint drill smaller than the final size of the hole is used and the hole is bored with a drill the same size as the centerpoint drill, it can then be enlarged to the desired size.

5.3 MOLDING CUTTING ON THE DRILL PRESS

A molding-cutting attachment is available for most drill presses. Figure 5.6 shows a molding-cutting attachment mounted on the drill press spindle in place of the chuck. The attachment consists of a shank with a spindle at the bottom of the shank with a bearing below it. A series of variable-diameter wheels can be mounted on the bearing below the cutter and the edge of the piece on which the molding is to be cut rides on the wheel

Fig. **5.5 A centerpoint drill.**

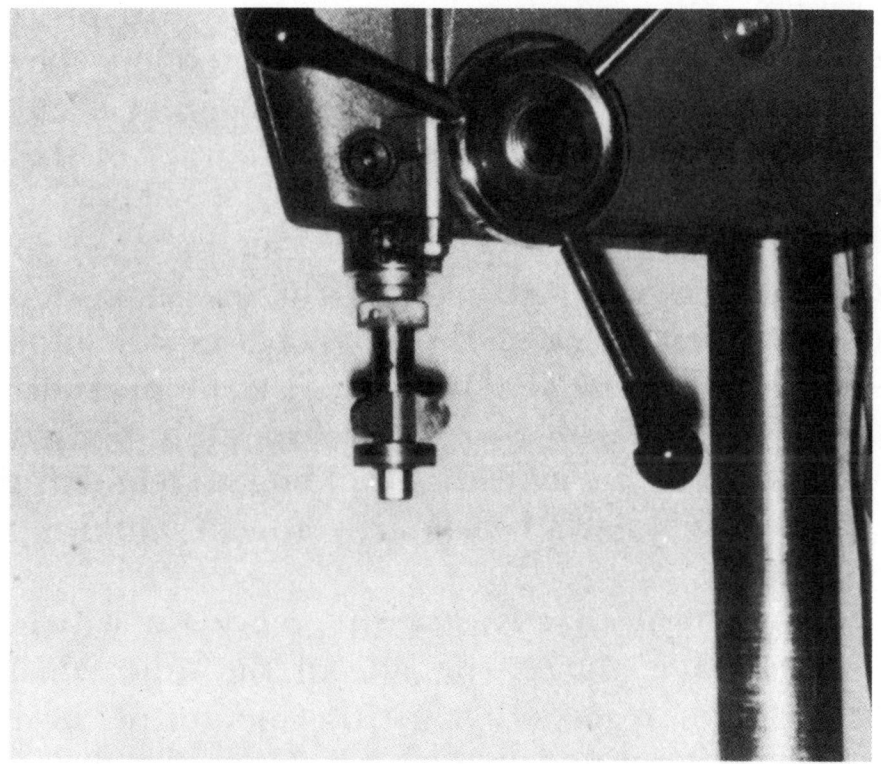

Fig. **5.6   A molding-cutter attachment on a drill press.**

below the molding cutter. The size of the wheel used depends on the depth from the edge of the piece to which the molding will be cut.

To cut a molding, the piece of lumber is placed on the drill press table and the spindle is lowered to a point where the bottom of the molding cutter will make a shallow cut and the spindle is locked in this position. With the drill press operating at high speed, the piece is pushed through, keeping the edge firmly against the wheel at the bottom. After the first cut is made, the spindle is lowered slightly and a second cut is made. This process is continued until the cut is as deep from the top of the stock as desired. The amount of cut made on each pass depends on the nature of the molding and the hardness of the wood. You can determine this by experiment, always remembering that in this process a considerable sidewise thrust is generated on the drill press spindle, and trying to make too great a cut can damage the spindle and destroy the accuracy of the drill press.

The drill press molding cutter can be used to cut moldings on curved pieces of nearly any shape as long as it is possible to keep the edge below the molding against the wheel at the bottom. A drill press with a molding-cutter attachment is therefore an important addition to the power saw with a molding attachment since the power saw is used primarily to cut moldings on straight pieces.

It is important to advance the piece on which the molding is being cut on the side of the unit, which allows the cutter to move against the motion of the piece of lumber as indicated in Fig. 5.7. Since the drill press spindle rotates clockwise, the piece of material being cut must be advanced on the *right*-hand side of the spindle. Any attempt to advance the piece on the left-hand side can lead to serious injury since the cutter will be likely to grab the piece and jerk it away from you, pulling your hands into the cutter blades.

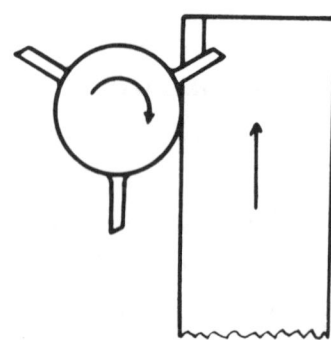

Fig. **5.7 The orientation of the cutter and the piece on which a molding is being cut.**

A number of different shapes of molding cutters are available for use with the drill press molding attachment. These cutters are the same as those used on a shaper. Figure 5.8 is a photograph of a molding being cut with a drill press molding cutter.

Fig. **5.8   A molding being cut.**

5.4   CUTTING MORTISES ON THE DRILL PRESS

The mortising tool is a device for boring a square hole. This is accomplished by means of a square chisel which surrounds a rotating wood bit. The wood bit, which projects slightly beyond the chisel, bores a round hole of diameter slightly less than the length of a side of the chisel. As the unit is forced down into the wood, the sharp edges of the chisel cut a square surrounding the round hole made by the bit, forcing the waste wood into the flutes of the bit. The waste wood is carried up to an opening on the side of the chisel where it is discharged. Figure 5.9 shows a cross section of the tool, where *A* is the chisel and *B* is the rotating bit.

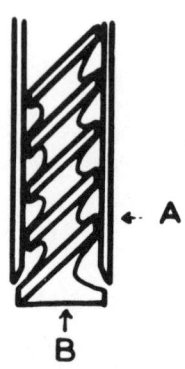

Fig. 5.9   Cross section of a mortising tool. *A* **is the chisel and** *B* **is the bit.**

In order for the mortising tool to function, it is necessary for the chisel, which does not rotate, to be raised and lowered with the boring bit, which rotates. The spindle on the drill press rotates within a tubular housing that does not rotate but which is raised and lowered with the spindle. A frame which supports the chisel is attached to the bottom of the spindle housing and when the chisel is mounted in place the bit is inserted up through the chisel and its shank is clamped in the chuck with its bottom end projecting slightly below the bottom of the chisel. It is usually necessary to remove the chuck while the frame for holding the chisel is being attached and, after it is attached, the chuck is replaced. Each size of chisel must be used with its corresponding size of boring bit. It is not necessary to have a large number of sizes of mortising chisels and bits, as it is possible to cut a rectangular hole larger than the tool by making a series of square cuts. I have a 1/4" and a 3/8" chisel and have been able to make any desired size of square and rectangular hole by making multiple cuts. When cutting a mortise with the smaller size mortising chisel, it is desirable to cut the square hole at each end of the mortise first and then make a series of cuts to remove the waste wood between them, because when a cut is made with no wood on one side of the chisel, there is some tendency for the chisel to be deflected slightly away from the wood so that the side of the hole will not be quite vertical. The depth of the mortise is controlled by properly setting the stop on the spindle.

The bit should be operated at relatively low speed and the unit must be firmly forced into the wood to make

Fig. **5.10 A mortising attachment on a drill press.**

the chisel cut the wood away from the sides of the hole; however, care must be exercised to avoid overheating the bit and chisel. To be certain that the mortise will be straight, it is helpful to clamp a temporary fence to the drill press table. Figure 5.10 is a photograph of a mortising attachment on a drill press.

5.5 THE DRILL PRESS VISE

When boring holes in small pieces of material, it is often difficult to hold them with your fingers. If the bit grabs in the hole, the piece may be jerked free and rotate with the bit, often resulting in bruised knuckles. A convenient device for holding small pieces of material on the drill press table is the drill press vise shown in Fig. 5.11. This type of vise can be obtained as a simple, general purpose vise which cannot be tilted on its base or as an angle vise which can be tilted on its base. The vise shown in Fig. 5.11 is an angle vise which can be

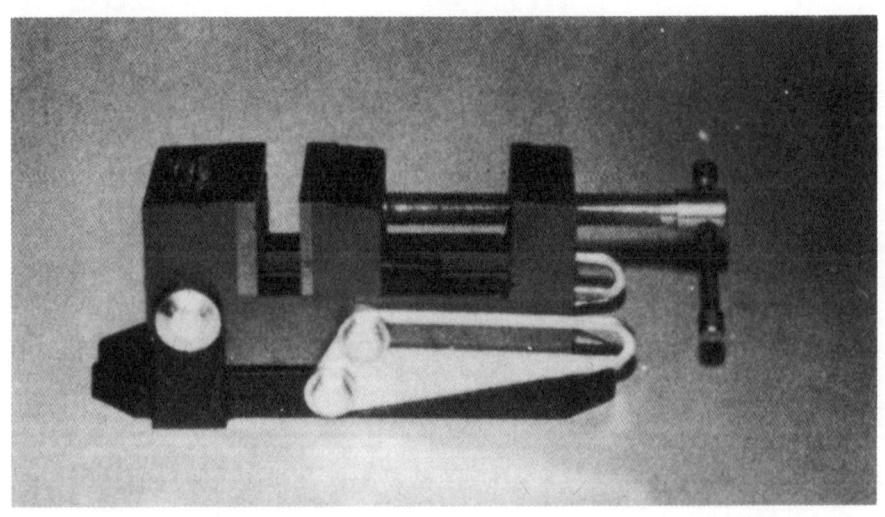

Fig. **5.11   A drill press vise.**

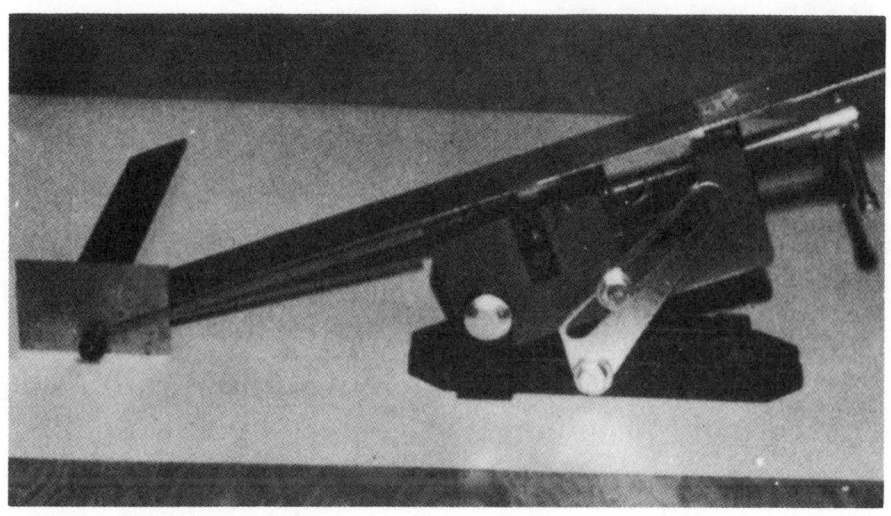

Fig. **5.12 Setting a drill press vise angle with a protractor and square.**

tilted to permit boring a hole at an angle. These vises are quite massive and they provide a large area to hold to, so the pieces being bored can be held without danger of jerking loose.

When boring at compound angles, such as is sometimes necessary when cutting mortises in chair legs, one of the angles can be set with the drill press table and the other set with the angle vise, or the drill press table can be left horizontal and the piece can be clamped at one of the angles between the jaws of the vise and the vise can be tilted to the other angle. The angle vise has a very small scale for reading the angle of tilt so its accuracy is inadequate for most work. When it is necessary to set the tilt to higher precision than this scale permits, it should be set by placing the vise on a flat surface, such as a saw table, and using a protractor to set the angle. Figure 5.12 shows an arrangement for setting the angle with a protractor and square.

5.6 HAND POWER BORING TOOLS

Hand electric drills are obtainable in three sizes, 1/4", 3/8" and 1/2". The size designation refers to the size of the chuck but the motor for a 1/4" drill is generally smaller than that for the larger sizes so that the units of smaller chuck size are usually smaller and lighter. The 3/8" size is the most popular since some types are not very much larger than the 1/4" models and there is some advantage in having the larger chuck available.

It is now possible to purchase 3/8" drills with variable speed motors. The motor is operated by a trigger in the handle and the speed is controlled by the amount of deflection of the trigger. It is even possible to purchase battery-operated hand electric drills. A storage battery is enclosed in the motor housing and, when the drill is not being used, it is connected to the battery charger.

It is important, when using hand electric drills operating from 110-volt or 220-volt power, to have the drill equipped with a three-wire cord with the ground wire properly grounded. Since it is inconvenient to have a long cord on the drill, you should have a 25' three-wire extension cord so that it will be possible to operate the drill at any location in your shop.

There are several accessories available for use with hand electric drills in addition to the standard drill bits. There is a disc sander and polisher attachment, a saber saw attachment (see Fig. 3.32), a small circular saw attachment that makes the drill a small hand power saw, and a vibrator sander attachment. It is also possible to purchase a speed-reducer kit. This kit includes a small speed-reducer transmission with a 1/4" shaft that fits into the drill chuck and a 1/2" chuck is attached to the other end. The transmission is a planetary type, so when the drill chuck is rotating and the bit is in the hole, the drill will not rotate but a drum on the outside of the transmission will rotate. By grasping the drum with your hand so that it stops, the bit will rotate at a lower speed than that of the small drill chuck. In this way you can control the operation of the large chuck by controlling the rotation of the drum. I have drilled 3/8" holes in masonry with a small 1/4" electric drill equipped with the speed reducer, using a special masonry bit. Also, the speed-reducer kits usually contain special large screwdriver bits and socket wrenches which can be held in the speed-reducer chuck. The hand control on the transmission drum is particularly important when these accessories are used.

5.7 SHARPENING CUTTING TOOLS USED WITH POWER BORING EQUIPMENT

There are few people with sufficient skill to freehand sharpen a twist drill. In order for the drill to cut properly, both cutting lips must make the same angle with the axis of the bit, both lips must simultaneously bottom in the hole, and the angle of relief on both lips should be the same. If you use twist drills to any great extent, a drill-grinding attachment for your grinder is a good investment. The cost of one of these attachments can be recovered by saving a few drill bits. The attachment mounts on the bench in a position such that the drill is ground on the side of the grinding wheel and, with the drill oriented so that one lip will be ground, it is advanced by means of a screw at the back of the attachment so that a small amount of metal will be ground away as the drill is swept across the grinder by rotating the attachment in its support. The drill is advanced in small steps and the process repeated until the entire surface of the lip is properly ground. The drill is turned over in the attachment and, without changing the screw setting, the attachment is gently rotated until the second lip contacts the grinding wheel. Do not apply much pressure and frequently turn it away from the wheel to allow the drill to cool. When the surface is ground sufficiently to permit the attachment to be swept completely around, the second lip will be ground exactly the same as the first. The drill must be placed in the attachment with the edge of the land against the small lip on the bottom of the channel in which the drill rests. You should carefully follow the instructions which accompany the attachment for its installation and use, for these will vary somewhat with different attachments. When the attachment is installed, it does not interfere with other use of the grinder and it is always there when a drill needs to be resharpened.

Molding cutters should be sharpened in the same manner as cutters for use in the molding head for a circular saw as described in Section 3.7.

The bits for mortising attachments are similar to auger bits except that they do not have a feed screw. They should be sharpened in the same manner as auger bits as described in Section 2.4.1. Figure 5.13 shows the method of sharpening the mortising chisels, where A is a fine file. The bevel is on the inside of the chisel and since the surface of the bevel is slightly curved, an oval pattern file works very well for this. Do not file the outside of the edge of the chisel, as this would cause it to cut a hole smaller than the upper part of the chisel and it would then bind in the hole.

Fig. **5.13   A cross section of a mortising chisel showing the use of a file *A* for sharpening the chisel.**

5.8   SAFETY

1. Never wear loose clothing which can catch on the chuck or drill bit when operating a power boring tool.
2. If you have long hair, put it up in a net or cap.
3. Never operate a power boring tool unless the bit is firmly clamped in the chuck with the chuck wrench. If the chuck turns on the bit shank, it can score the shank so the bit will not run true.
4. Do not get in the habit of leaving the chuck wrench in the chuck, as you may forget to remove it before starting the drill. You can be injured by a flying chuck wrench.
5. Always be certain that the piece of material being bored is firmly supported so it cannot rotate with the bit if it grabs in the hole.
6. Always use a three-wire cord to connect the tool to the power circuit and be certain that the ground wire is properly connected.

6

TURNING TOOLS

6.1 THE LATHE

In order to do any turning, it is necessary to have a machine to rotate the work piece while various chisels are applied to remove wood symmetrically about the axis of rotation. Such a machine is called a lathe. The minimum requirements for a lathe for wood turning is a headstock or spindle, capable of being rotated by a motor, with a means of attaching the work piece so that the spindle will cause it to rotate; a tailstock to support the free end of the work piece, when that is necessary, with a means of providing a bearing on which the end of the work piece can rotate; and a bed to connect the headstock to the tailstock and to provide a means of supporting a tool rest for the various turning chisels.

A lathe which provides the bare minimum of requirements can be used only for wood turning but entirely satisfactory wood turning can be performed with it. Such a lathe, with motor and bench, can be purchased for as little as $100. More sophisticated lathes provide more different turning speeds, more different methods of attaching the work piece to the spindle and more elaborate beds.

I do metal turning as well as wood turning so I have a

lathe which is considerably more sophisticated than the simplest wood-turning lathe. Figure 6.1 is a photograph of my toolmaker's lathe. Since the spindle in the head-stock of the lathe must turn the work piece, a means must be provided for attaching it to the spindle. In many in-stances a spur is used which will turn the work piece when it is mounted between the spur and the tailstock center. When a work piece cannot be mounted against a tailstock center, it is necessary to attach it firmly to the headstock spindle. A special metal plate, to which the work piece can be attached with screws, can be mounted on the spindle or, with lathes similar to that shown in Fig. 6.1, a three-jaw or four-jaw chuck may be used. Figure 6.2 shows a series of units for attaching the work piece to the spindle, where A is a spur, B is a plate with a screw at the center, C is a small three-jaw chuck, D is a large three-jaw chuck and E is a four-jaw chuck with independent jaws.

Fig. **6.1   A toolmaker's lathe.**

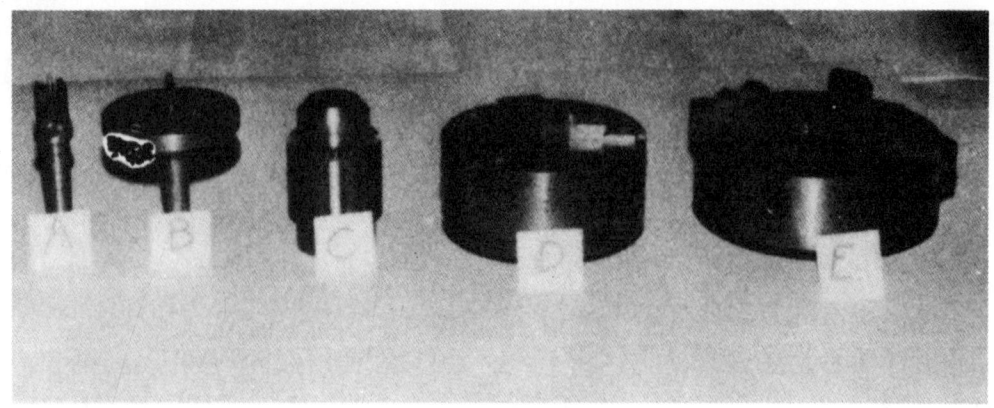

Fig. **6.2 Units for attaching a work piece to the headstock spindle.** *A* **is a spur,** *B* **is a plate with a screw at its center,** *C* **is a small three-jaw chuck,** *D* **is a large three-jaw chuck and** *E* **is a four-jaw chuck.**

The jaws on three-jaw chucks work together so that the opening is always on the axis. Three-jaw chucks are suitable for attaching cylindrical work pieces and four-jaw chucks should be used for attaching square work-pieces to the spindle.

When the end of the work piece must be supported by the tailstock, the tailstock center serves as the bearing on which it turns. The tapered hole formed by a center-point drill (see Fig. 5.5) has the same taper as the lathe tailstock center. This provides a good bearing surface on which the work piece can rotate.

The lathe bed on simple wood-turning lathes consists of one or two steel tubes on which the tailstock can slide and be positioned for the length of the work piece. It also serves to hold the tool rest which supports the chisels used in the turning. The lathe bed on a lathe of the type shown in Fig. 6.1 is fabricated of cast iron and the top surfaces are machined so that the tailstock can be moved on a set of ways and clamped in any position. The tool holder moves on a separate set of ways and can be driven parallel to the axis of the work piece either manually or by power. When hand chisels are used, a steady rest, similar to that used on simple wood turning lathes, is mounted on the tool holder, but cutting tools similar to those used in metal turning may also be mounted in the tool holder.

6.2 WOOD TURNING CHISELS

A set of wood-turning chisels usually contains about four to six different shapes and each shape of chisel is available in several different widths. The shapes are illustrated in Fig. 6.3 and they consist of the gouge, the skew, the parting tool and the round-nose. In addition, a square nose and a spear point may also be added. Fig-

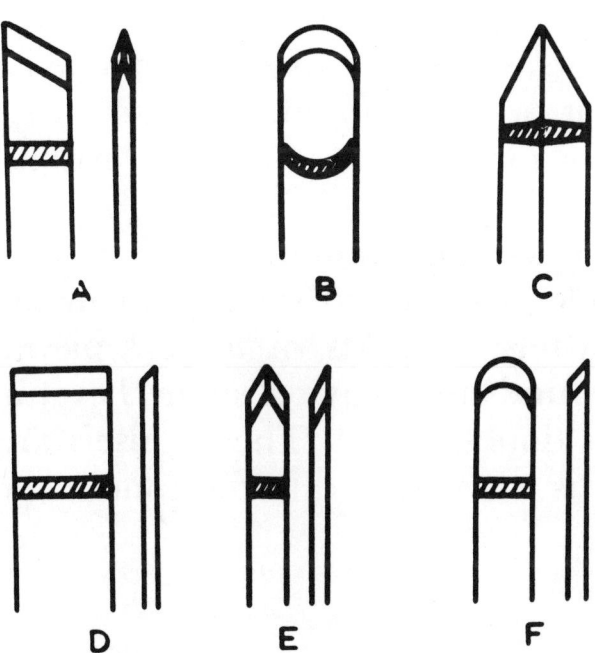

Fig. **6.3** **Shapes of standard chisels used in wood turning.** *A* **is a skew,** *B* **is a gouge,** *C* **is a parting tool,** *D* **is a square nose,** *E* **is a spear point and** *F* **is a round nose.**

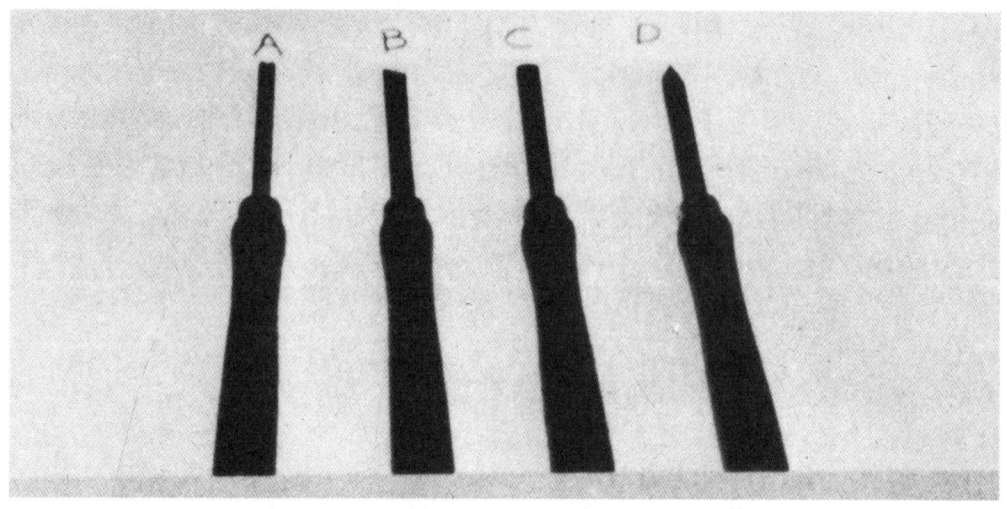

Fig. **6.4   A set of wood-turning chisels.** *A* **is a gouge,** *B* **is a skew,** *C* **is a round nose and** *D* **is a parting tool.**

ure 6.4 is a photograph of a set of wood-turning chisels consisting of a gouge, a skew, a round-nose chisel and a parting tool.

The gouge is hollow, like an ordinary woodcarvers gouge, but it is round nosed. It is used for roughing out work and for making cove cuts in a turning. The skew is double ground and flat. It is used for smoothing cylinders and for cutting shoulders and beads. The parting tool is double ground and is used for making sizing cuts and for cutting off parts of a work piece. The square nose and round nose have a single bevel and the spear point has a double bevel. These chisels are used where their shape is best adapted to the contour being cut.

6.3   TURNING BETWEEN CENTERS

In order to illustrate the use of wood-turning tools in

the various steps in wood turning, I will turn various contours on a piece of scrap stock. We will assume that this piece is to be a small table leg with a square portion at the top. Such a piece will normally have the diameters of the largest diameter portions of the turning equal to a width of the square portion of the leg. It will be necessary, therefore, to carefully square the piece before mounting it in the lathe. After the piece is squared, it should be cut to the proper length with the ends square relative to the sides and the ends should be marked as indicated in Fig. 6.5.

Fig. 6.5   The method of marking the ends of a piece to be turned.

The end which is to be the bottom of the leg should have a hole bored with a centerpoint drill, 1/4" or larger, at the intersection of the lines *a* and *b*. The position of the center of this hole should be carefully marked with a center punch to insure the accuracy of location of this hole and the hole should not be bored beyond the taper on the centerpoint drill.

The end of the piece, which will be the top of the leg, will be driven by the headstock spindle. If a spur (*A* in Fig. 6.2) is used, a hole of diameter equal to that of the point on the spur should be bored at the intersection of the lines *a* and *b* and shallow kerfs should be cut with a back saw exactly on the lines *a* and *b*. The driving jaws on the spur will be engaged by these kerfs. If the lathe has a four-jaw chuck, the chuck can be used to hold the square end of the piece but it will be necessary to accurately center the end of the piece in the chuck. When the work piece is mounted in the lathe, a drop of oil should be placed on the tailstock center.

In the first step, the leg below the square portion is turned down to a cylinder of diameter such that the four flat portions just disappear. With a wood-turning lathe, this is done with the gouge and steady rest. Figure 6.6 shows how the chisel is held to perform this operation. There will be a shoulder at the bottom of the square section. To start this, the work should be nicked with the skew on edge as indicated in Fig. 6.7. With the parting tool the shoulder is rough cut, as indicated in Fig. 6.8, and it is finished with the skew, as indicated in Fig. 6.9.

Fig. **6.6 Turning the work piece down to a cylinder with a gouge.**

Fig. **6.7** **Knicking the shoulder with the skew on edge.**

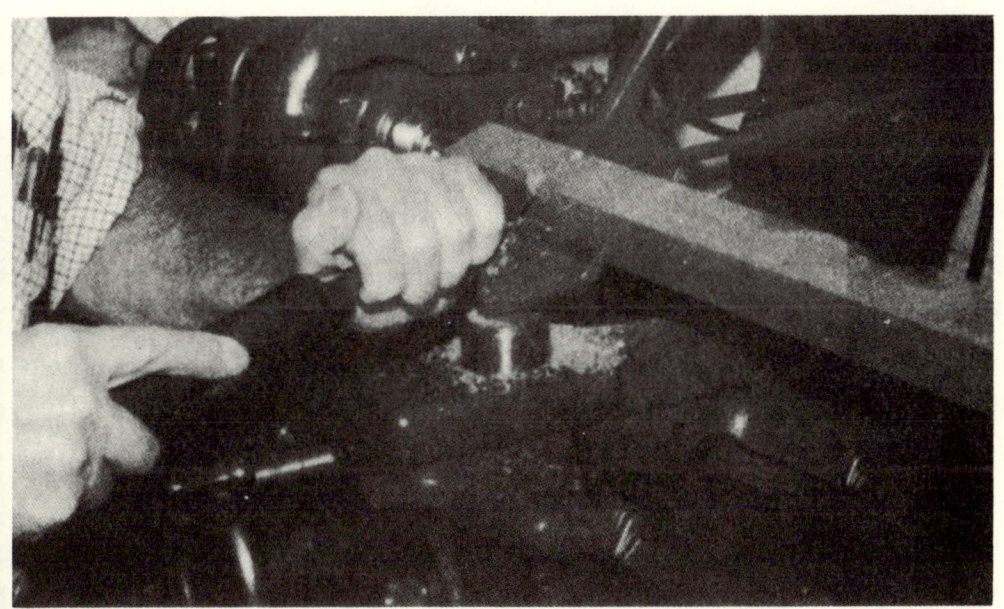

Fig. **6.8** **Rough cutting the shoulder with the parting tool.**

Fig. **6.9 Finishing the shoulder with the skew.**

With a lathe like that shown in Fig. 6.1, the piece can be turned down to a cylinder using a round-nosed metal-cutting tool, as indicated in Fig. 6.10. The power drive for the tool holder can be used, so little effort is needed for this operation. When it is necessary to form a shoulder, it is desirable to form it with wood-turning tools, even though the metal-turning tools are used to form the remainder of the cylinder.

We are now ready to start turning a design on our piece. The first step is to mark the positions of the various elements of the design. To illustrate the use of the various chisels, we will lay out the design illustrated in Fig. 6.11 on our cylinder. The positions of the portions *a, b, c, d, e* and *f* are marked on the cylinder, as indicated in Fig. 6.12. These are turned with the parting tool, to the desired diameter. Figure 6.13 shows the use of the parting tool to turn these portions and Fig. 6.14 shows how the diameters are checked with the calipers.

Fig. **6.10** **Turning the cylinder with the metal-cutting tool.**

Fig. **6.11** **A practice design.**

Fig. **6.12** **Marking the positions of** *a, b, c, d, e* **and** *f* **from Fig. 6.11 on the cylinder.**

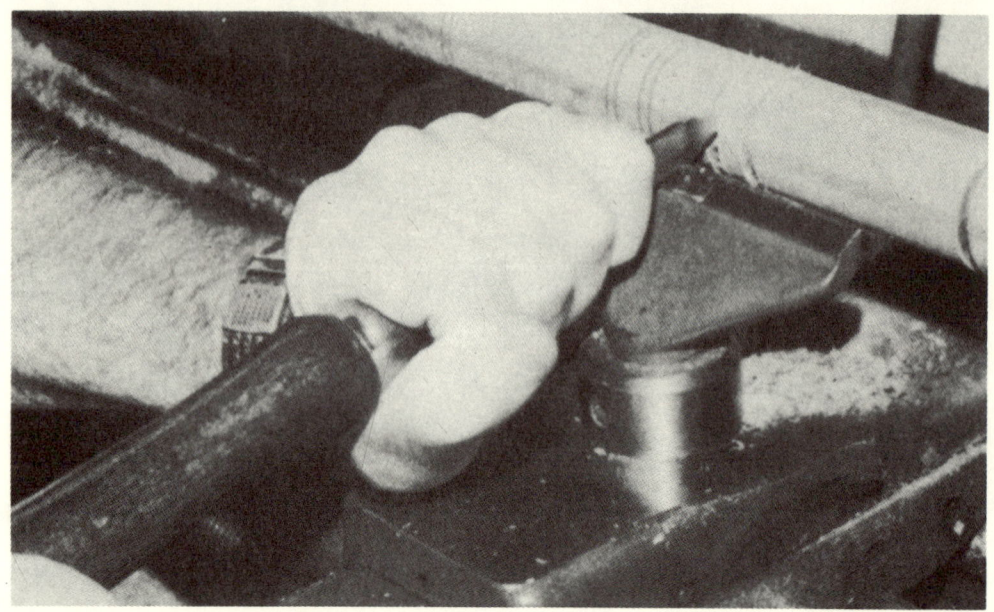

Fig. **6.13** **Turning the points** *a, b, c, d, e* **and** *f* **with the parting tool.**

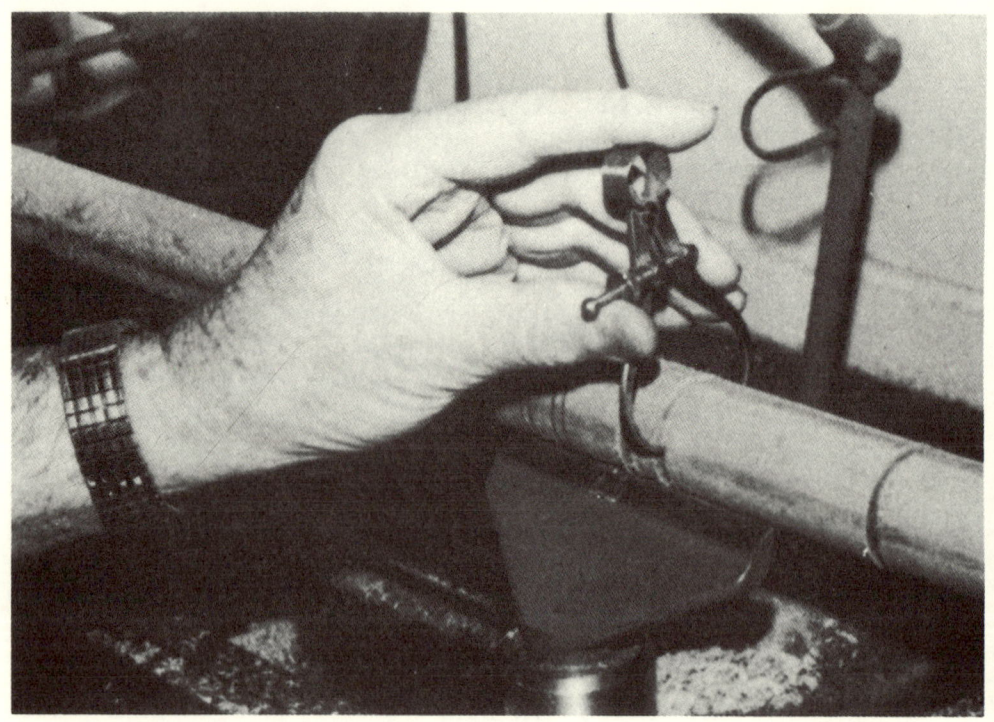

Fig. **6.14** **Checking the diameter with a caliper.**

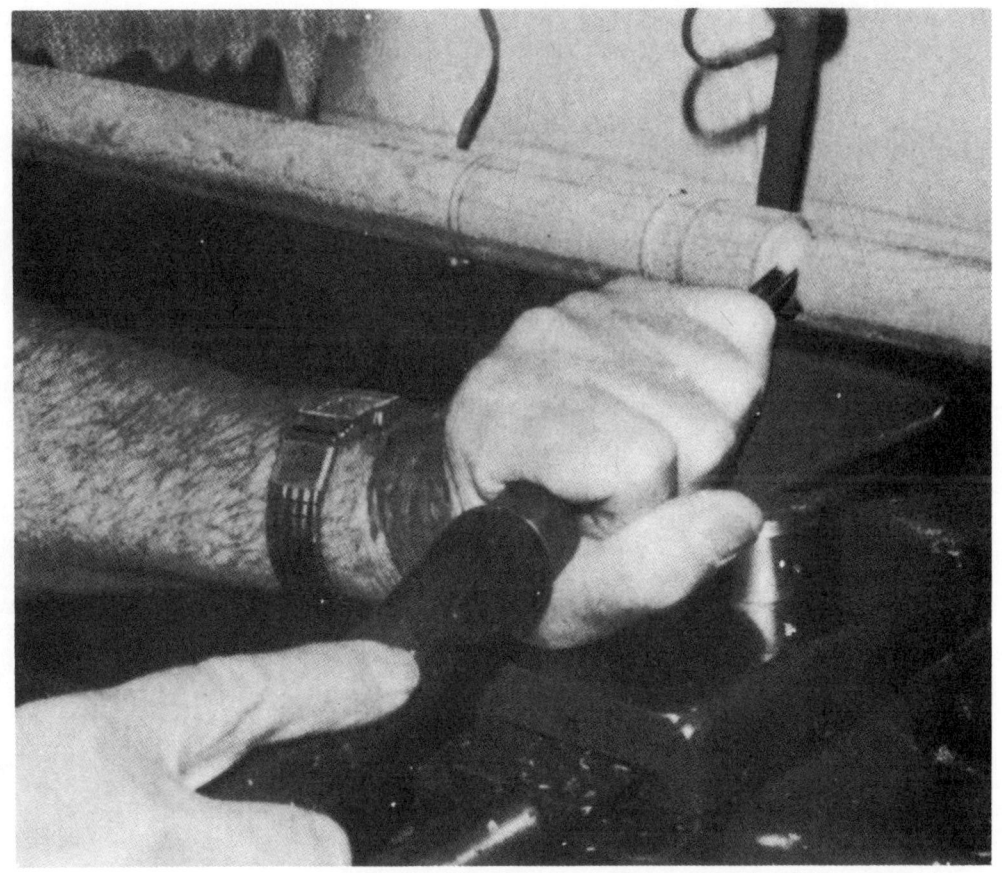

Fig. **6.15**   **Turning the coves with the round nose.**

Figure 6.15 shows the turning of the coves *g* and *h* with the gouge. Again the caliper is set for the minimum diameter of the cove and is used to check the diameters of the coves.

Figure 6.16 shows the turning of the knob *i* between the coves with the skew. The portions *j* and *k* are turned with the round-nose chisel and finished with the skew. The portions at the ends are turned with the skew, although the spear point can be used to advantage here if you have one available.

If you have never done any wood turning, a practice sample such as this is a good means of becoming familiar with the use of the various wood-turning tools.

Fig. **6.16   Turning the knob with the skew.**

There may be instances where you will wish to turn half-columns for use in decorations on a piece of furniture such as a clock. It is not very practical to turn a column and rip it with a saw because of the waste due to the thickness of the saw blade. A better way to do this is to glue two pieces of wood together with paper between them and the centers can then be located at the glued junction. After the piece is turned, it is possible to separate the two pieces at the paper junction so that two full half-columns will result.

After the pieces are turned it will be necessary to smooth them. I like to use a file for the first step in the smoothing and then finish with sandpaper. The sandpaper should be cut in narrow strips. Do not use turning chisels or files on the wood after sandpaper has been used because the sandpaper will leave some grit in the surface of the wood which will dull a tool or a file. Never use a file without a handle on a rotating piece. When you are sanding a piece in the lathe, you should cover the bed with some paper towels to keep the sandpaper grit from falling on the bed of the lathe.

6.4 TURNING PIECES NOT SUPPORTED BY A TAIL-STOCK CENTER

A piece which cannot be supported by a tailstock center may be a disc with a hole through its axis, such as a base for a candlestick. Such an item may be supported by a bolt which can be held in the lathe chuck. When the piece is supported in this manner while it is being turned, the piece will be symmetrical about the axial hole.

A more common kind of piece which is turned without a tailstock center is a bowl. Before starting to turn a bowl, you should cut an outside and inside pattern of cardboard as indicated in Fig. 6.17, where A is the pattern for the outside and B is the pattern for the inside of the bowl.

You should start with a piece of lumber with at least one flat side and which is at least as thick as the outside height of the bowl. You can mark a point which will be the center on a flat side of the piece and with a compass point at the center you can scribe the outside circumference. A circular disc can be cut out on a band saw, cutting slightly outside the mark. Now bore a hole of diameter equal to the diameter at the bottom of the threads of holder B in Fig. 6.2 into the flat side of the

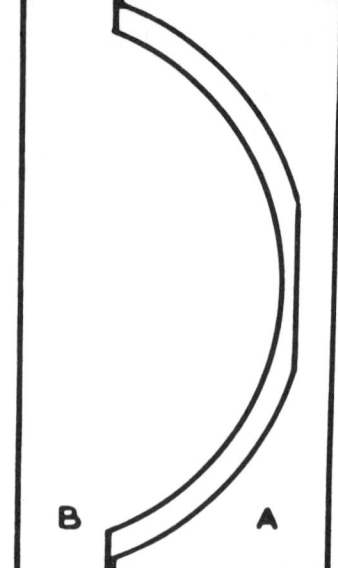

Fig. **6.17   A pattern for turning a bowl.**

Fig. **6.18   A wood disc set up for turning the outside circumference.**

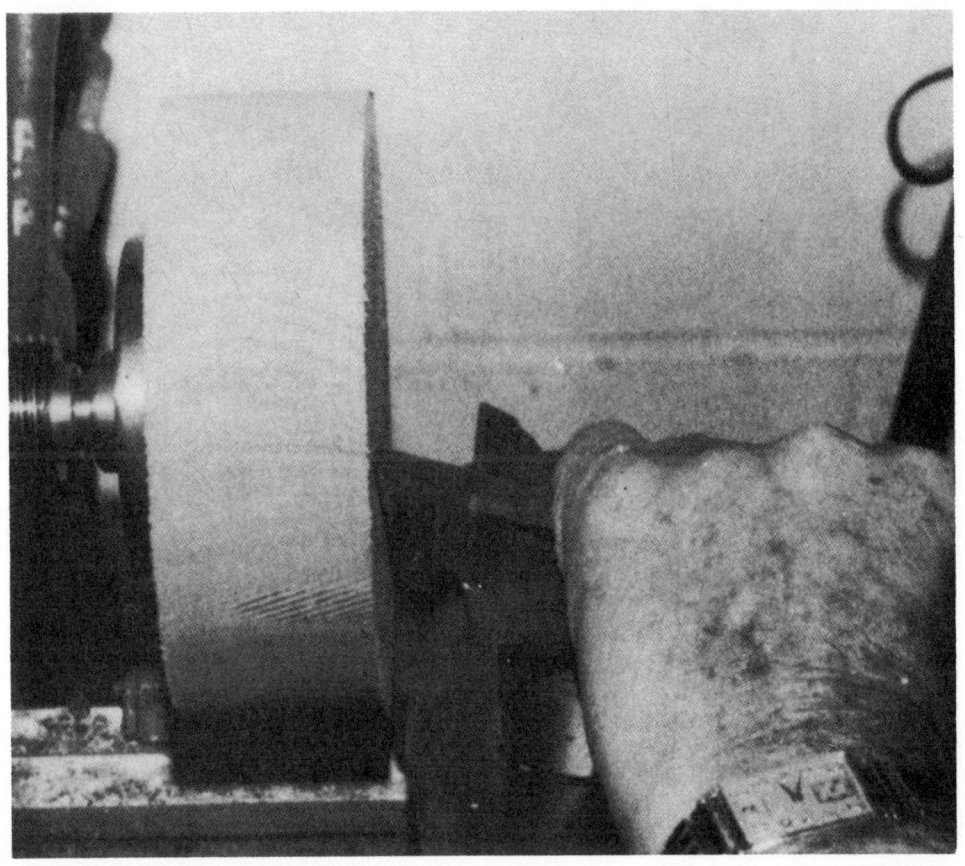

Fig. **6.19** **Arrangement for turning the flat bottom with the round nose and steady rest.**

disc, as deep as the length of the screw. Screw the holder in place until the plate is firmly against the flat surface of the wood disc. The unit can now be mounted on the spindle of the lathe.

You should first turn the rim of the disc to the proper diameter. You are then ready to shape the outside. Figure 6.18 shows the wood disc at this stage mounted on the lathe spindle. The first step is to turn the outside surface flat. This may be done with the round-nose chisel and steady rest as indicated in Fig. 6.19, finished with the skew or flat nose chisel, or, with a metal turning lathe, the tool arrangement shown in Fig. 6.20 can be used.

Fig. **6.20   Arrangement for turning the flat bottom with the metal-turning tool.**

When the bottom surface has been turned flat and parallel to the top surface, the circumference of the flat portion of the bottom can be marked as indicated in Fig. 6.21. The outside contour can now be turned, using the gouge for the rough turning and finishing with the round-nose chisel and the skew. Figure 6.22 shows the turning in process. The contour is shaped so that the pattern *A* in Fig. 6.17 fits it. When the contour has been properly shaped, the surface can be smoothed with a file and sandpaper.

In order to turn the inside of the bowl it is necessary to devise a means of holding it. This can be done by

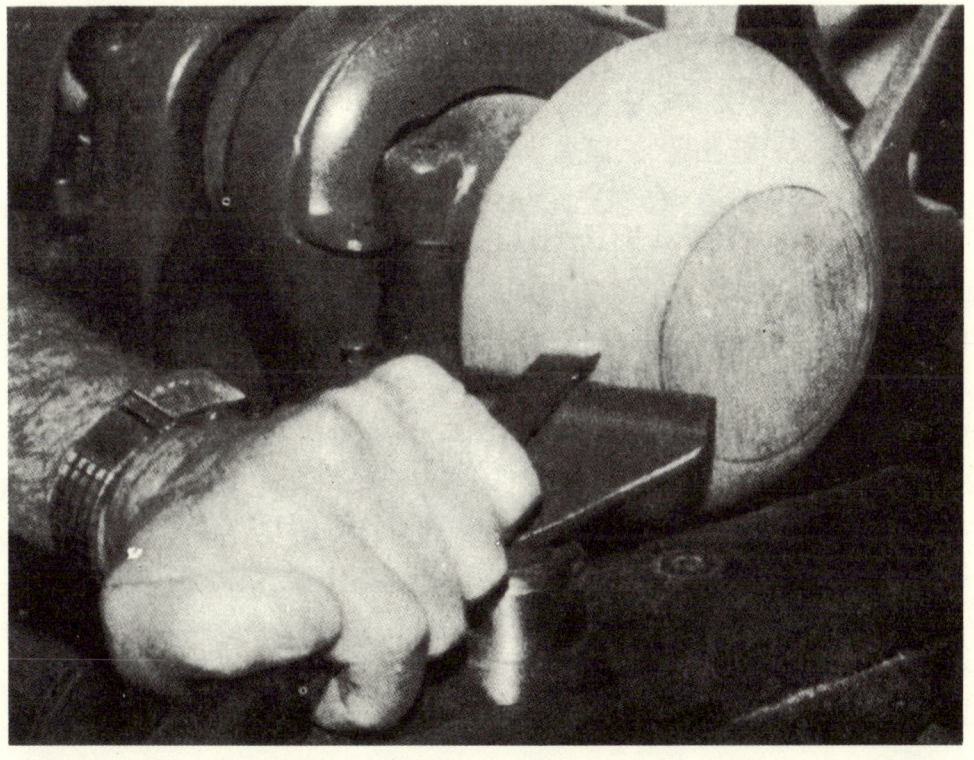

Fig. **6.21** **Marking the circumference of the flat portion of the bottom.**

Fig. **6.22** **Turning the curved outside with the skew.**

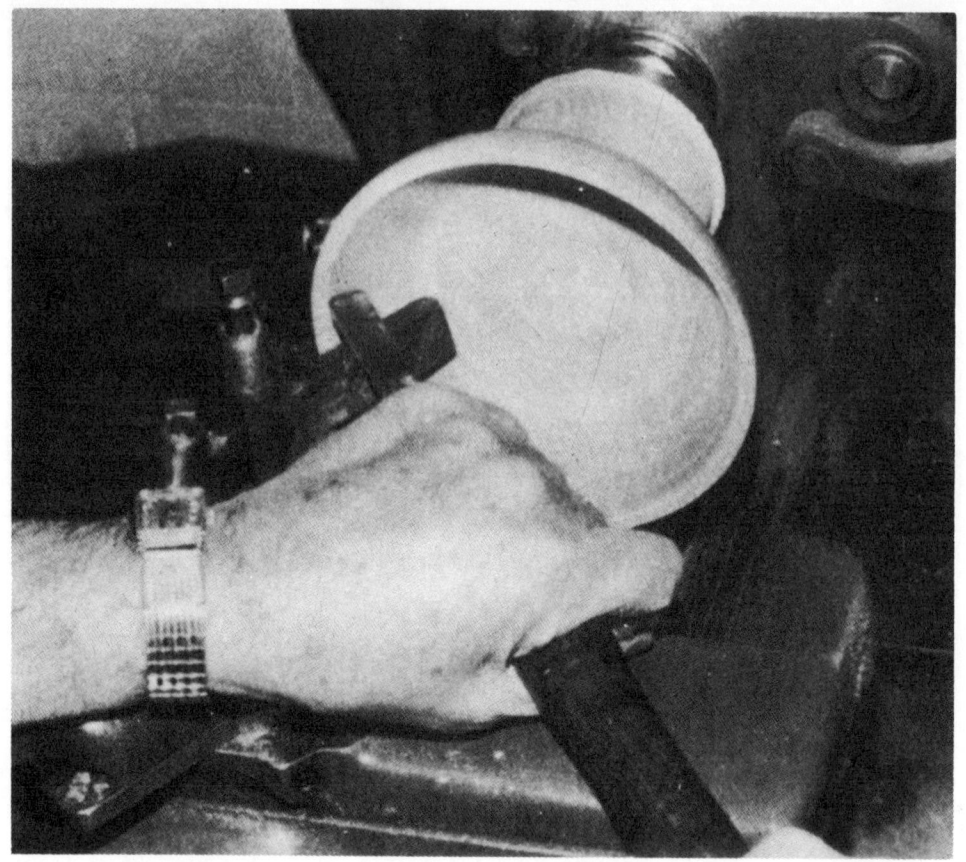

Fig. **6.23** **Turning the inside with the round nose.**

turning a cylindrical disc the same diameter as the flat bottom of the bowl and boring a hole in the center to accept the screw on the plate *B* in Fig. 6.2 The bowl can be glued to the holding disc with the disc centered on the flat portion of the bottom with paper between the bowl and the holding disc so that they can be easily separated later.

Figure 6.23 shows the inside of the bowl being turned with the round-nose chisel. Wood is removed until the pattern *B* in Fig. 6.17 fits the contour. The space between *A* and *B* in Fig. 6.17 determines the thickness of the wood in the bowl. The inside surface can be smoothed with sandpaper.

6.5 SHARPENING WOOD TURNING TOOLS

The gouge and round-nose chisel are sharpened in the same manner as wood carving gouges as described in Section 2.5.1, except that, the round-nose chisel being flat, the flat side is honed in the same manner as the unground side of a chisel or plane iron. The flat nose is sharpened in the same manner as the chisel, as described in Section 2.5.1.

The skew and spear point are ground on the flat side of the wheel as indicated in Fig. 6.24. These tools are

Fig. **6.24   Method of grinding a skew or a spear point.**

ground on both sides and are honed in the same manner as a chisel is honed.

The parting tool is ground on the two edges as indicated in Fig. 6.25. It should be ground so that the sharp edge will be on the center line.

The turning tools should be honed frequently, because it is important for them to be sharp in order to do satisfactory work.

6.6 SAFETY

Wood turning can be hazardous if proper precautions are not observed.

Fig. **6.25 Method of grinding a parting tool.**

1. Do not wear loose clothing when operating a lathe. You should wear a shirt with short sleeves.
2. Do not wear a necktie.
3. If you have long hair, put it up and wear a cap.
4. When you use a file to smooth the work, never use a file without a handle. When files are purchased, they do not have handles, but they have a tang to which a handle can be attached.
5. Before starting the lathe, be certain that the workpiece is properly mounted and that there is adequate clearance for the workpiece to turn.
6. You should wear safety glasses when roughing out work and when grinding chisels.
7. Keep the tool rest close to the work and remove it when sanding or filing the workpiece.
8. Always keep the cutting tools sharp and hold the hand tools firmly when they are being used.
9. Do not attempt to use your woodcarving tools for wood turning. They are not sufficiently strong to withstand the forces involved in wood turning.

7

THE POWER ROUTER

7.1 INTRODUCTION

The power router is not found in the amateur's shop as often as the other power tools. It consists of a motor in a cylindrical housing with a collet-type chuck at the lower end of the motor shaft for holding the various types of cutter bits. The motor mounts in a frame that terminates on a circular plastic disc at the bottom, with a hole at the center through which the cutter bit can extend. The motor can be raised or lowered in the frame, usually by means of a rack and pinion, and there is a scale for indicating the level of the cutter bit. A means is usually provided for setting the zero of the scale so that the actual depth of the cut can be read directly, usually to 1/64".

Cutter bits are available for cutting V-grooves, veins (round bottom grooves) and dadoes (square bottom grooves). Various shapes of molding cutters are also available for cutting edge moldings and rabbeting cutters are available for cutting rabbets on an edge.

An edge guide is used when veins, V-grooves or dadoes are to be cut parallel to a straight edge. A special attachment is used with the edge guide to permit cutting veins, V-grooves or dadoes at a fixed distance from a curved edge.

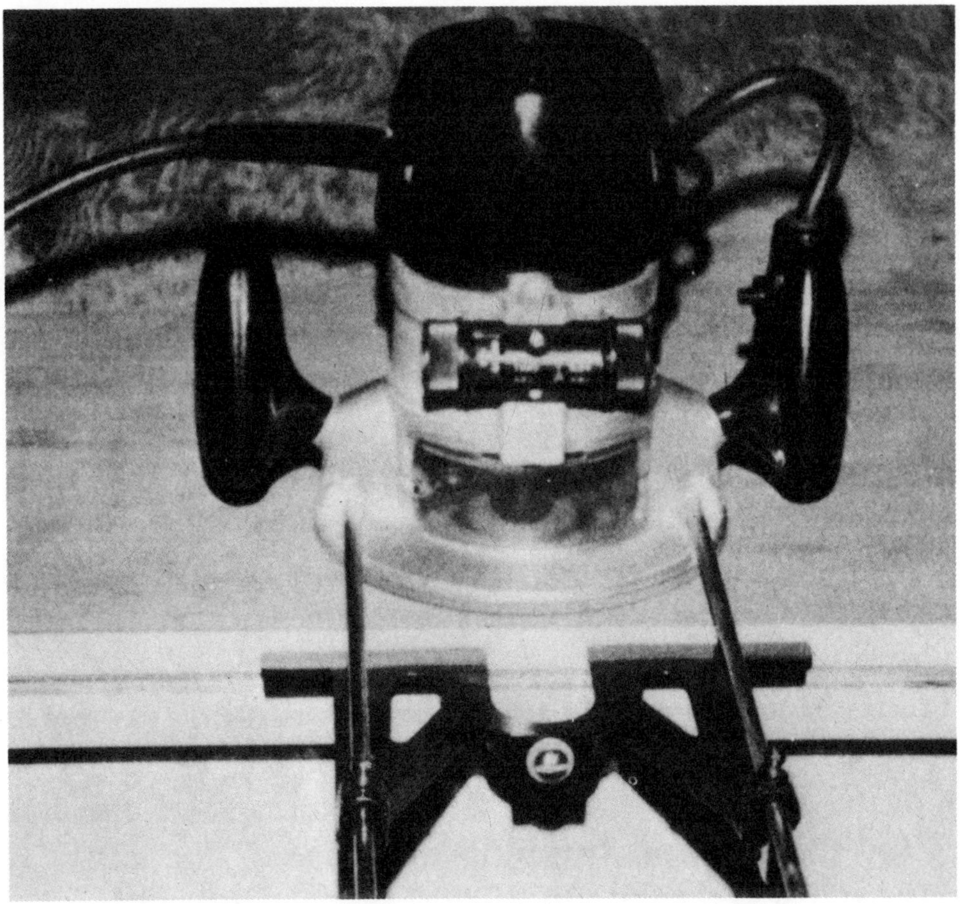

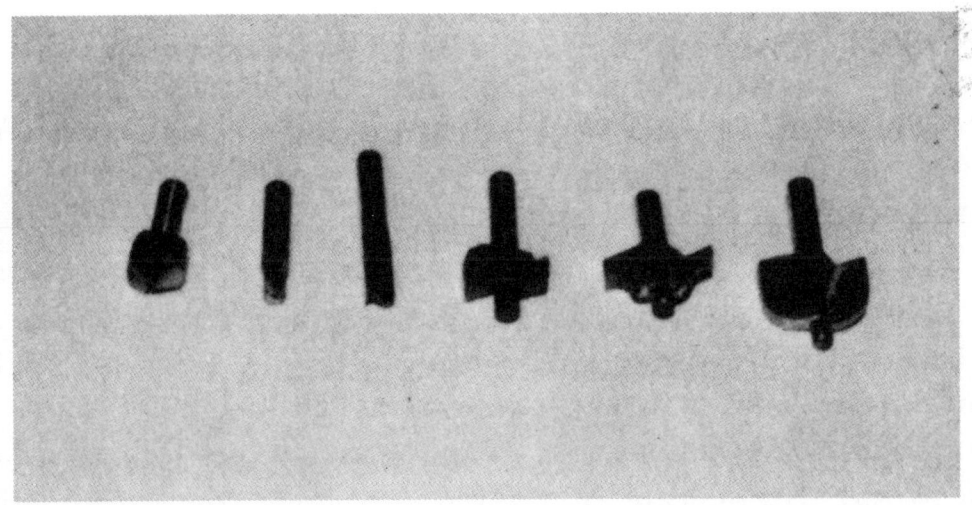

Fig. **7.1   A power router with the edge guide attached.**

Fig. **7.2   Some cutter bits for use in a power router. From left to right; a V-groove cutter, a veining cutter, a dado cutter, a rabbeting cutter and two edge molding cutters.**

Figure 7.1 shows a small power router with the edge guide in place. Figure 7.2 shows a V-groove cutter, a veining cutter, a dado cutter, a rabbeting cutter and some molding cutters. Various widths of veining cutters and dado cutters are available and dadoes wider than the dado cutter can be cut by making successive passes with different settings of the edge guide. The edge molding cutters and the dado cutters usually have a pilot at the bottom that rides against the edge of the workpiece below the cut. Some dado and molding cutters have detachable pilots so that they can be used with or without the pilot. When the pilot is not used, the edge guide must be used.

There is a pair of handles on the router frame that can be grasped for guiding the machine through the cut. A trigger-type switch is located on one of the handles for starting and stopping the motor. The cutter bit rotates at a speed of about 20,000 revolutions per minute.

7.2 CUTTING VEINS, V-GROOVES AND DADOES

Figure 7.3 shows a router being used to cut a vein. A dado and a V-groove have previously been cut with different settings of the edge guide.

The rate of feed of the cutter into the workpiece is important. If the rate of feed is too great, it is necessary for the cutter to take large bites into the wood and, if they are too large, the chips will be knocked off rather than cleanly cut, resulting in splintering and gouging of the wood. If the rate of feed is too slow, the cutter will bounce around in the cut, causing the sides to be rippled and sometimes resulting in burning of the tool.

Fig. **7.3** **A vein being cut with the router. Note the parallel V-groove and dado which were previously cut with different settings of the edge guide.**

Cuts made with V-groove cutters and small veining and dado cutters can usually be made to the full depth of the cut in one pass, particularly in soft wood. With large cutters, in hard wood, the depth of cut should usually be limited to not more than 1/8" in a single pass.

The motor and cutter on a router revolve in a clockwise direction so the reaction on the handles of the router will be counterclockwise. This reaction is not great in normal cutting but, if a hard spot in the wood, such as a knot, is encountered, this reaction can result in a sudden kick. Because of this, the handles of the router should always be firmly grasped to insure that the cut will be straight. A template guide bushing can be mounted in the spindle hole of the base plate to follow guides of various shapes in routing. For this type of cutting, the tool travel should be as indicated in Fig. 7.4.

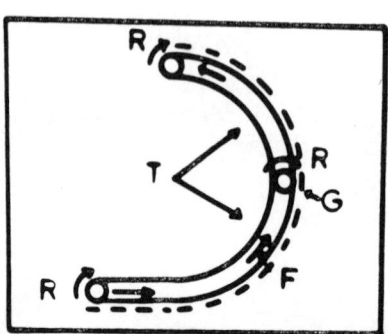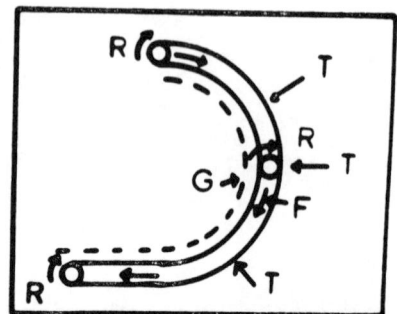

Fig. **7.4   The relation between direction of feed, direction of rotation of the cutter and direction of thrust applied to the handles relative to a template guide.** R **indicates the direction of rotation of the cutter,** T **indicates the direction of thrust and** F **indicates the direction of feed. The dashed** $-G$ **indicates the template guide.**

7.3   CUTTING RABBETS OR MOLDINGS

Rabbets and moldings are cut on the edges of the workpiece. If the cutter has a pilot at the bottom end, the pilot can serve as a guide by riding on the edge of the workpiece below the cut. However, if the cutter does not have a pilot, it will be necessary to use the edge guide to control the depth of the cut from the edge of the workpiece.

It is usually not possible to cut a full rabbet with a single pass of the cutter. If the cutter has a pilot, it is most convenient to make the first pass at the full depth of the cutter from the edge of the workpiece but at less

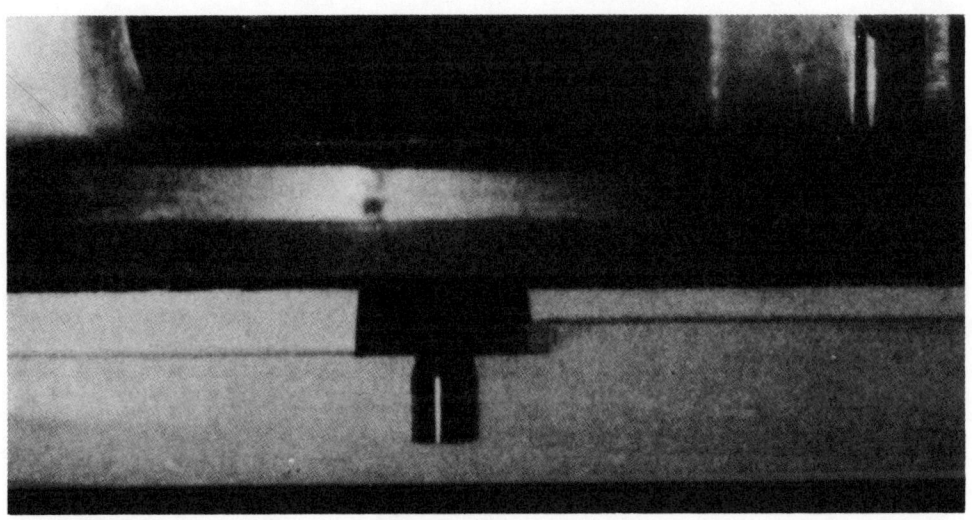

Fig. **7.5   Cutting a rabbet with a rabbet cutter with a pilot.**

than the full depth from the top. The depth of the cut from the top can be increased with successive passes of the cutter by lowering the motor in the frame. Figure 7.5 is a photograph showing a rabbet cutter with a pilot after one pass has been made and while a second pass is being made.

If the rabbet cutter has a pilot, it can be used only for cutting rabbets having a depth from the edge equal to or less than the width from the pilot to the cutting edge. If a rabbet of greater depth from the edge is to be cut, a cutter without a pilot should be used and the depth from the edge should be controlled with the edge guide.

Various edge molding cutters are available and they are used in a manner similar to that of the rabbet cutter. Usually these cutters have a pilot and the molding is cut by making a series of passes with the depth of the cutter increased between successive passes. Figure 7.6 is a photograph of a cutter in the process of cutting a molding on the edge of a workpiece. Figure 7.7 shows the direction of feed and thrust with a router used for edging on inside and outside curves.

Fig. **7.6   Cutting an edge molding.**

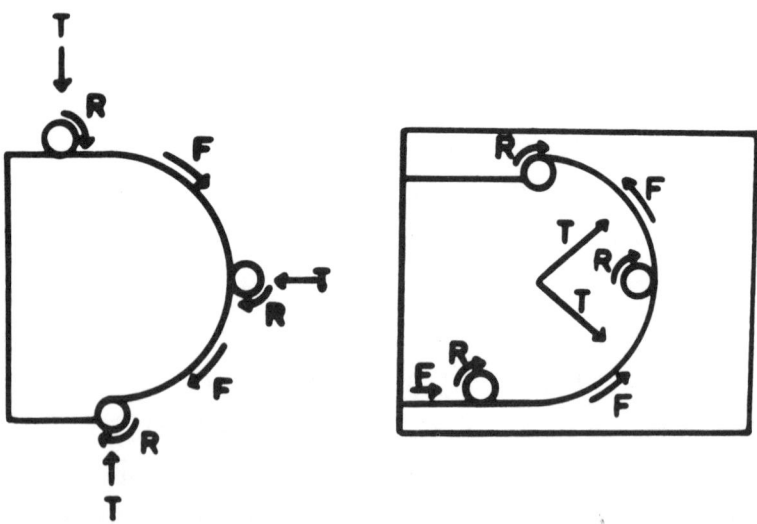

Fig. **7.7   The relation between direction of feed, direction of rotation of the cutter and direction of thrust when edging inside or outside edges. *R* indicates the direction of rotation of the cutter, *T* indicates the direction of thrust and *F* indicates the direction of feed.**

7.4 ACCESSORIES FOR USE WITH A ROUTER

There are a number of accessories for use with a power router. These include:

1. A template which can be used for routing dovetail joints for drawers.
2. A template for routing the countersinks for butt hinges.
3. A template set of numbers and letters with a holder to permit mortising the characters in the wood.
4. A special guide and bit for trimming formica on the edges of table tops and counter tops.
5. A pantagraph for use in reproducing designs into wood engravings.

You should consult your router manual and the instructions provided with the various accessories for the proper methods of using them.

7.5 CONVERTING THE POWER ROUTER TO A SMALL SHAPER

It is difficult to control the router when making cuts on small pieces of lumber. This problem can be made easier by making a shaper table in which the router is used as the shaper mechanism. A piece of 3/4" fir plywood about 2' by 4' can serve as the table. The base plate should be removed from the router and a special base plate, made from a piece of 1/4" hardboard about 11" or 12" square, should be prepared. This special base plate should have a hole in the center the same size as the spindle hole in the standard base plate and screw holes should be bored corresponding to the screw holes in the standard base plate and they should be counter-sunk from the smooth side of the hardboard.

Cut a square hole at the center of the plywood about 2" less on a side than the dimensions of the hardboard and carefully mark the outline of the hardboard base on the plywood so that a rabbet can be cut around the square hole in the plywood to permit mounting the hardboard base flush with the surface of the table. You can use your router with a dado cutter to cut this countersink. Now mount the hardboard base in the countersink in the plywood table with countersunk wood screws at each corner.

A fence can be constructed from a straight piece of 3/4" lumber about 3" wide and long enough so that it can be pivoted near one end of the table and clamped with a C-clamp at the other end. This fence performs

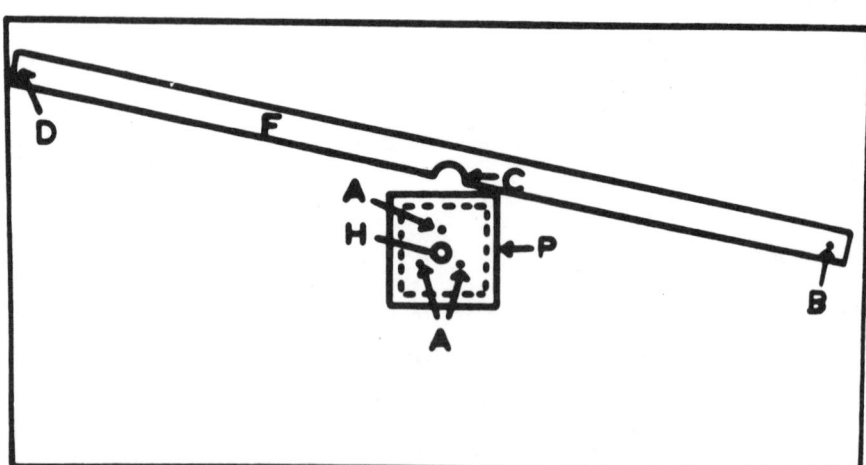

Fig. **7.8   A shaper table to permit using a router as a shaper.** *P* **is the hardboard base plate,** *H* **is the spindle hole, the holes** *A* **are for the router mounting screws,** *F* **is the fence,** *C* **is the cutter clearance in the fence,** *B* **is the pivot for the fence and** *D* **is the end of the fence which must be clamped to the table with a C-clamp when the fence is used.**

the same function as the edge guide when the router is used in the normal manner.

Figure 7.8 shows schematically how the shaper table is made. This is a very useful arrangement. The table can be mounted on legs to provide a permanent tool in the shop or, to conserve space, it can be put aside and temporarily mounted when it is to be used.

When using this arrangement, remember that the router is inverted so that the rotation of the cutter bit will be counterclockwise. The operations shown in Figs. 7.4 and 7.7 will thus be reversed.

7.6 CUTTING SMALL MOLDINGS

In order to safely cut a small molding, it should be cut on the edge of a larger piece of lumber. After the molding is cut it can be sawed off. Figure 7.9 shows a piece of lumber on which a small molding has been cut and

Fig. **7.9 A small molding showing the saw cuts to separate it from the parent piece of lumber.**

the two saw cuts necessary to separate the molding from the parent piece.

7.7 SHARPENING ROUTER CUTTER BITS

Router molding cutters should be sharpened in the same manner as molding cutters used on a saw or drill press (see section 3.7). In order to sharpen the other cutters it is necessary to use a special sharpening jig which provides a special holder for the tool and a small grinding wheel which mounts in the router chuck. You should follow the instructions provided with the sharpening attachment for your particular router.

7.8 SAFETY

1. Always be certain that the arbor of the cutter bit is securely clamped in the chuck with at least 1/2" of the arbor in the chuck.
2. Be certain that the screws holding the base to the router frame are tight.
3. Always use a three-wire cord to connect the motor to a properly grounded power receptacle.
4. When using the router as a hand-operated router, be certain that the workpiece is securely clamped.
5. The bit should be held clear of the wood until after the motor is started.
6. Always grasp the handles firmly to properly control the router.
7. When the cut is completed, turn off the motor and wait until it stops before moving the router.
8. When using the shaper table, use the fence for al straight cuts and be certain that it is securely clamped.

9. Maintain a 4" margin of safety when using the fence.
10. Wear safety glasses when using the router or shaper.

8

SANDING

8.1 INTRODUCTION

Sandpaper was originally a product prepared by gluing sand to a paper backing. The modern product consists of abrasives such as aluminum oxide, Carborundum or garnet with either paper or cloth as a backing material. The abrasive is applied electrostatically to the backing material so that the points of the abrasive point up. Although the name is obsolete, we still speak of this product as sandpaper.

Modern sandpapers are graded according to the size of the abrasive particles. Grades 50 through 100 are coarse, 180 through 280 are medium to fine and grades from 320 to 400 are very fine and suitable only for very hard wood. A good selection of grades to have available is 50, 100, 180, 200 and 280.

Paper-backed abrasives are less expensive than those with cloth backing. Never buy cheap sandpaper. Such a product is made from inferior abrasives and inferior backing. Some sandpaper is labeled "wet-or-dry." Such sandpaper can be used on surfaces wetted with oil, which is sometimes desirable.

8.2 HAND SANDING

In furniture construction, most of the surfaces to be sanded are either flat or round. We have already discussed the sanding of round surfaces turned on a lathe in sections 6.3 and 6.4. The flat portions to be sanded will vary greatly in width but, regardless of the width, it is desirable to have a uniform application of the sandpaper to retain a flat surface. The most primitive holder of the sandpaper is the operator's thumb or fingers. There are many small areas where this is the most convenient method of use of the abrasive but, due to the fact that the thumb and fingers are round, it is difficult to maintain a flat surface if an appreciable amount of wood must be removed with the sandpaper. The use of the thumb or fingers can quickly result in blisters because of the heat generated in the process. Because of the above difficulties, the sandpaper is often cut in strips and wrapped around wood blocks. The sizes of the blocks used will depend on the areas to be sanded.

Figure 8.1 shows the type of sandpaper holder that I prefer. It is made of rubber, which is not as rigid as wood and is, therefore, a more desirable support for the sandpaper than a wood block. The rubber holder takes a piece of sandpaper about 2 5/8" wide, of length equal to the length of the short side of a standard sheet of sandpaper. The strip of sandpaper is firmly held by three pointed pins in each of the slots on the ends of the rubber block.

The length of the portion of the strip on the bottom of the block is about 4 3/4", leaving about 1 3/4" on each end which might be considered a waste. However, when the portion on the bottom of the block is worn out, I

Fig. **8.1** **Rubber sandpaper holder.**

tear off these unused ends and save them for use with
small wood blocks or with my fingers for sanding in
small, restricted areas.

When sanding irregular surfaces, such as curved
areas, it is necessary to use the technique which woodcarvers use in smoothing their work. This consists of
using narrow strips of sandpaper that are held against
the surface with the thumb of one hand and pulled
through with the other hand. To avoid blisters, the
thumb should be protected with a thin pad held in place
with adhesive tape. The thumb is too wide to hold the
sandpaper when sanding very narrow irregularities but
you can whittle properly shaped wood blocks from soft
wood to hold the sandpaper in place while it is pulled
through. When sanding narrow irregularities, cloth-
backed abrasive is superior to the paper-backed variety,
as the paper will break under the severe bending that is
required before the abrasive is worn out.

You should select the grade of sandpaper to use in
the first step in sanding according to the degree of

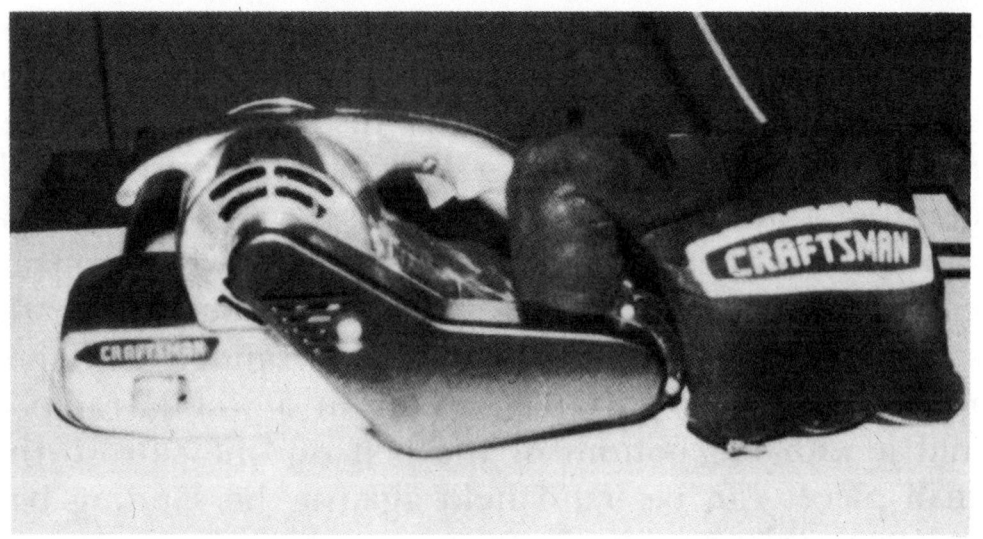

Fig. **8.2   A portable power sander.**

roughness to be smoothed out. If the wood is badly chipped, you may need to start with an abrasive grade as coarse as 50. This will leave scratches which must be removed with successively finer grades of abrasive. In most instances, the wood will be smooth enough so you can start with grade 100 or 120 abrasive and finish with about 240. I usually work in three steps, starting with grade 120, then going to grade 180, and finishing with grade 240. You should always sand parallel to the grain. When smoothing end grain, the surface should be filed or planed to remove torn fibers before the sanding is started.

8.3   THE POWER SANDER

The abrasive for a power sander is applied to an endless cloth belt that runs over two rollers, one of which is driven by a motor. Figure 8.2 shows a portable power sander. Such sanders are made in two sizes. The larger size uses belts 4" wide and the smaller size uses belts 3" wide. There are larger belt sanders made for permanent mounting in the shop. These sanders often have a large disc to which circular sandpaper discs can be cemented in addition to the sanding belt. These permanently mounted sanders are quite expensive and are seldom found in the amateur's shop. They are, however, convenient for sanding small pieces which can be held in the operator's hands while they are being sanded. It is possible to mount a portable sander in a wood frame to hold it with the bottom of the belt on one side so that small pieces can be hand held against the sanding belt. It is possible to purchase a portable power sander equipped with a dust bag and a vacuum blower to suck the dust into the bag. This is an important accessory, since the dust from a sander can be a serious nuisance in the shop.

In using a portable power sander, you should not exert any pressure beyond the weight of the sander. You should constantly keep the machine in motion, otherwise the belt will dig grooves in the piece being sanded. Figure 8.3 illustrates a desirable pattern of motion to follow in sanding a large area such as a table top.

Generally the sanding is done with the belt running parallel to the grain of the wood. There are instances where a glued-up piece is uneven. In section 2.3 it was indicated that in such instances you should plane the piece perpendicular to the grain depending on the bottom of the plane to bridge the low areas. In the same way, you can do the initial sanding with the power sander perpendicular to the grain depending on the plate over which the sanding belt runs on the bottom of the sander, to bridge the low places. When the irregularities have been sanded out, the piece should be sanded parallel to the grain to remove the scratches left by the cross-grain sanding.

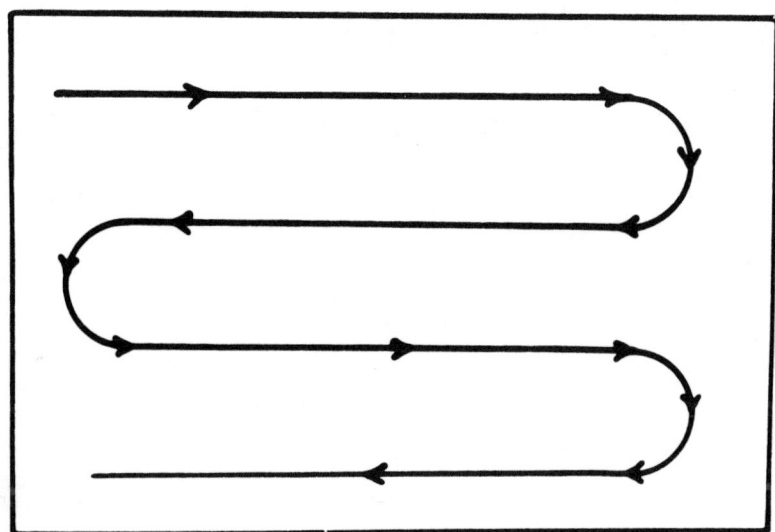

Fig. **8.3** **Pattern of motion of a power sander over the work piece.**

There are three grades of abrasive on sanding belts; fine, medium and coarse. You will seldom have need for the coarse abrasive belts and the medium grade belts should be used only on fairly rough pieces. The fine belts have considerably coarser abrasive than sandpaper used in hand sanding. The high speed of the sanding belt allows a coarser abrasive to be used. When the sanding is completed with the power sander, the piece should be hand sanded with grade 180 and 240 abrasive to obtain a finished surface.

8.4 ABRASIVE WHEELS

Abrasive wheels are available with Carborundum particles welded to the surface of an iron disc that can be mounted in place of a saw blade on a table saw. These wheels usually have a coarse abrasive on one side and a fine abrasive on the other. The workpiece should be lightly pressed against the abrasive wheel to avoid overheating the wood. The heat developed causes the resin in the wood to deposit on the abrasive wheel. This deposit can be removed with commercial "gum and pitch remover," which can also be used for removing resins deposited on saw blades.

9

SHOP TALK

This book will not make the reader an instant craftsman. Craftsmanship can come only as a result of practice. However, practice without guidance may be largely wasted effort and can result in damaged tools and serious accidents.

I have attemped to show the methods of use, adjustment, and maintenance of the tools which may be found in the amateur's shop so that he can develop the necessary skills to do fine work.

It is impossible to overemphasize the importance of properly sharpened tools. Even a skilled craftsman cannot do good work with dull tools and dull tools are more conducive to accidents than are sharp tools.

I do not favor equipping a shop with a full complement of power tools immediately. I believe that it is better to start by purchasing some good hand tools and learning how to use them. You will need a bench grinder to keep your planes, chisels, and gouges sharp and your first projects should be small, simple pieces.

When you have mastered the use of hand tools, purchase a power saw. I prefer a table saw rather than a radial arm saw because, even though the radial arm saw appears more glamorous than the table saw, you can work to higher precision with the table saw.

When you have mastered the power saw, the order in which you add other power tools will depend on your interests.

As you develop your skill, your enjoyment will increase. There is great satisfaction in being able to construct beautiful and useful things from wood.

About fifty years ago, I took a course in woodworking and, when I completed it, I felt that I knew quite a lot about it. Since then I have been making furniture and repairing old furniture as a hobby. I find that I am still learning new things about wood and how to work with it. This is the exciting thing about woodworking. Each project is a new challenge and a large part of the fun comes from the thrill of solving the problems that come up.

GLOSSARY

abrasive — Any substance used to grind, wear down or smooth a surface.

anvil — A steel-faced iron block or machine part on which metal is shaped.

arbor — A metal shaft or axis on which a revolving cutting tool is mounted.

arkansas stone — A superior variety of novaculite found in Arkansas and used for fine honing of tools.

auger bit — A tool for boring holes in wood, consisting of a shank with spiral channels that end in two spurs; for marking the outline of the hole and a tapered feed screw.

awl — A pointed instrument for marking surfaces or punching small holes.

back saw — A short, fine-toothed saw stiffened with a metal rib along its back edge.

band saw — A saw on which the cutting teeth are on a steel band which runs over two wheels.

bit
A replaceable part of a compound tool that actually performs the function of cutting.

block plane
A small plane having the iron set at a lower pitch than other planes and used chiefly on end grains of wood.

brace
A crank-shaped instrument for holding and turning auger bits.

caliper
A measuring instrument having two legs or jaws that can be adjusted to determine thickness, diameter or distance between surfaces.

Carborundum
A commercially produced silicon carbide.

C-clamp
A C-shaped general purpose clamp.

centerpoint drill
A small twist drill used to make centers in a piece of work.

center punch
A machinist's hand punch, consisting of a short steel bar with a hardened conical point at one end.

chipbreaker
A plate attached to the top of a plane cutting-iron to break up the chips.

chisel
A tool consisting of a short metal bar with a sharpened edge at one end.

chuck
An attachment for holding a workpiece or tool in a machine.

collet	A casing or socket for holding a drill or other tool.
combination saw	A saw blade that can be used for either crosscut or rip sawing.
combination square	A tool consisting of a steel rule that slides through a protractor head which can be secured at any point on the rule.
compass	An instrument for scribing circles.
cove	A molding with a concave cross section.
crosscut saw	A saw designed chiefly to cut across the grain of wood.
dado	A groove with straight sides.
divider	An instrument for measuring or marking and transferring dimensions.
dog	A clamp for communicating motion.
dovetail cutting saw	A small back saw.
dovetail joint	A flaring tenon and a mortise into which it fits to make an interlocking joint.
dowel	A headless smooth pin.
dowel center	A metal piece with a thin shoulder and a prong at the center that can be placed in a dowel hole and used for marking the center of a dowel hole in an adjoining piece of material.

doweling jig	A jig used to locate dowel pin holes.
dowel pin joint	A joint reinforced with dowels.
drill press	An upright drilling machine.
expansion bit	A bit with a cutting blade that can be adjusted to various sizes.
fence	An attachment to a woodworking machine that controls the location or extent of the cut.
flexible curve	A piece of flexible material that can take various forms; used as an aid in drawing.
flute	A groove of curved section.
fore plane	A plane, about 18" long, used for planing long pieces or large areas.
French curve	A curved piece of flat material used as an aid in drawing.
gauge	An instrument for measuring a particular dimension of an object.
gouge	A chisel with a concave-convex cross section.
grain	The fibrous structure of wood.
grinder	A machine for sharpening tools with abrasive wheels.
headstock	A part of a lathe that holds the revolving spindle and its attachments.

hone	To sharpen with a fine grit stone.
idler	A wheel or roller used to transfer motion.
index	Something that serves as a pointer or indicator.
jack plane	A plane, about 14" long, used for general purpose planing.
jig	A device used to maintain mechanically the correct positional relationship between a piece of work and the tool working on it or between parts of the work during assembly.
jig saw	A machine saw with a narrow reciprocating blade.
jointer	A power planer for smoothing wood.
jointing	The act or process of making a joint. To file down saw teeth to the proper height.
kerf	A slit or notch made by cutting with a saw.
land	The metal between the flutes of a twist drill.
lathe	A machine in which work is rotated about a horizontal axis and shaped by a fixed cutting or boring tool.
lathe bed	The portion of the lathe which supports the headstock, tailstock and tool holders.
mallet	A hammer that has a cylindrical, barrel-shaped head of wood or other soft material.

marking gauge	A tool for scribing a line parallel to the edge of a piece of work.
mill file	A single-cut, tapered or blunt file.
miter	A surface forming the beveled end or edge of a piece of material.
miter box	A device for guiding a hand saw at the proper angle in making a miter joint.
miter gauge	A tool with graduations used to set a saw at any desired angle.
molding head	A holder for mounting molding cutters on a machine.
mortise	A hole, groove or slot into which some other part fits.
mortise and tenon joint	A joint formed by a tenon fitted into a mortise.
parting tool	A narrow cutting tool used primarily for cutting off sections of work on a lathe.
pawl	A pivoted tongue on one part of a machine that is adapted to fall into notches or interdential spaces on another part.
pilot	A cylindrical projection on the end of a tool used to guide it.
plane	A tool for smoothing or shaping a surface of wood.

plywood
: A structural material consisting of sheets of wood glued together with the grains of adjacent layers at right angles.

protractor
: An instrument for laying out or measuring angles.

quill
: A hollow shaft surrounding another shaft and used in various mechanical devices.

rabbet
: A channel or groove cut out of the edge or face of any body.

radial arm saw
: A saw with the mechanism mounted on an arm above the table.

ratchet
: A mechanism composed of a ratchet wheel and pawl.

resawing
: The process of sawing lumber edgewise to produce thinner lumber.

rip saw
: A saw designed chiefly to cut in the direction of the grain of wood.

round nose
: A flat wood turning chisel with a rounded cutting edge.

router
: A machine with a rapidly revolving spindle and cutter for milling out the surface of wood.

saber saw
: A portable electric jigsaw.

sanding	Smoothing or polishing with sandpaper.
sandpaper	A paper with abrasive cemented to its surface, used for smoothing or polishing.
saw set	An instrument used to give set to saw teeth.
scribe	To mark by cutting or scratching a line.
scroll saw	A thin hand saw for cutting curves or irregular designs.
skew	A flat chisel, used in wood turning, which has its cutting edge at an angle relative to its length.
skirt	A decorative piece on furniture, connecting the legs along the lower edge of a table top, chair seat or base.
sliding "T" bevel	An instrument similar to a try square with the blade adjustable in angle and free to slide in the handle.
smooth plane	A plane, about 9" long, for planing short pieces of lumber parallel to the grain.
spindle	A revolving piece.
spur	The circumferential cutter on an auger bit.
square	To form right angles. An instrument for indicating a right angle.
steady rest	A rest in a lathe for supporting a tool.

steel square	A carpenter's square made of steel.
stone	A conglomerate of abrasive for honing a tool.
tailstock	The adjustable or sliding head of a lathe containing the dead center.
tang	A piece that forms an extension from the blade or analogous part of a tool.
tapered saw	A saw blade that is thicker at the teeth than in the body of the blade.
tenon	A projection on a piece of wood for insertion into a mortise to make a joint.
tenon jig	A jig for use in cutting tenons on a saw.
try square	An instrument consisting of two straight edges secured at right angles to each other, used for laying off right angles and testing whether work is square.
twist drill	A drill having one or two deep helical grooves extending from the point to the smooth portion of the shank.
vise	A tool with two jaws for holding work.
way	A longitudinal guide or guiding surface on the bed of a machine.

INDEX

All numbers in italics refer to illustrations.